Marcio Jose Rodrigues Amorim

Synthesis and Characterization of Cerium-doped PVDF Samples

Marcio Jose Rodrigues Amorim

Synthesis and Characterization of Cerium-doped PVDF Samples

Electrical and optical characteristics of cerium-doped PVDF samples

ScienciaScripts

Imprint
Any brand names and product names mentioned in this book are subject to trademark, brand or patent protection and are trademarks or registered trademarks of their respective holders. The use of brand names, product names, common names, trade names, product descriptions etc. even without a particular marking in this work is in no way to be construed to mean that such names may be regarded as unrestricted in respect of trademark and brand protection legislation and could thus be used by anyone.

Cover image: www.ingimage.com

This book is a translation from the original published under ISBN 978-613-9-61541-4.

Publisher:
Sciencia Scripts
is a trademark of
Dodo Books Indian Ocean Ltd. and OmniScriptum S.R.L publishing group

120 High Road, East Finchley, London, N2 9ED, United Kingdom
Str. Armeneasca 28/1, office 1, Chisinau MD-2012, Republic of Moldova, Europe
Printed at: see last page
ISBN: 978-620-7-86185-9

Copyright © Marcio Jose Rodrigues Amorim
Copyright © 2024 Dodo Books Indian Ocean Ltd. and OmniScriptum S.R.L publishing group

Table of contents:

SYNTHESIS AND CHARACTERIZATION OF POLY(VINYLIDENE FLUORIDE) SAMPLES DOPED WITH CERIUM.

Marcio José Rodrigues Amorim

My wife Elizangela and my daughter Gabriely.

ACKNOWLEDGMENTS

To my advisor, Professor Evaristo Alexandre Falcäo, for his friendship, teachings and, above all, for his professional and personal example.

To my co-supervisor, Professor Eriton Rodrigo Botero, for his teachings and encouragement throughout the work.

To my laboratory and Master's degree colleagues, Laís, Bruna, Gustavo Ruivo, Gustavo Kern, Marileide, Carlos, Carlise, Cássia, Érica, Fernando, Haroldo, Keila, Lis, Rafael and Thiago, for their friendship, patience, encouragement, support and collaboration throughout this period.

To my family, especially my parents, José and Maria, my siblings, Marcos and Joice, my wife Elizangela and my daughter Gabriely for all their love and encouragement during all the stages of my life.

To the laboratory technician William Falco, for his friendship and help in carrying out various experiments.

To all the professors of the Chemistry Graduate Program, especially Professors Cláudio, Nelson and Andrelson for their teachings throughout this work.

To all my teachers during my undergraduate studies, especially Edemar Benedetti and Sandro Miguzzi.

To my great friends Robson Andrade, Flávio César, Nilton César and Matheus Sobreira for their encouragement and long conversations at difficult times in my life and in this work.

To my cousins Fátima and Jorge for their friendship and for their example of struggle during difficult times in their lives.

To my cousin José Luiz Matana (*in memorian*), for his struggle and will to live, but who is now at God's side.

To Professor Ivair Aparecido dos Santos for the dielectric measurements carried out on the samples.

To Laís Weber, who often helped me with sample preparation and data collection.

To my fellow undergraduates, Bruno Gabriel, Fabrício Aguero, Daniel Mendes, William Goulart and Carlos Amorim, for their friendship and coexistence over several years of our lives.

To the Lord my God, who in His infinite goodness and wisdom has allowed me to be part of His "Great Project" called life. Author of life, wisdom and science, He designed me, created me, loved me and allowed me to carry out this work by enabling me, granting me health, inspiring me and strengthening me at every moment.50

"The opposite of a correct statement is a false statement. But the opposite of a profound truth can be another profound truth".

Niels Bohr

SUMMARY

With the aim of combining the properties of cerium with those of PVDF, this work presents the synthesis and characterization of composites in the form of films, with potential for optical applications, carrying the properties of these two classes of materials. PVDF films were prepared in such a way as to combine the properties of cerium in the Ce^{3+} oxidation state (cerium chloride) and in the Ce^{4+} oxidation state (cerium sulphate). PVDF films doped with cerium sulphate ($Ce(SO_4)_2.4H_2O$) were prepared at the following concentrations: 0.0, 0.2, 0.4, 0.6, 0.8 and 1.0 %. PVDF films doped with cerium chloride ($PVDF/CeCl_3.7H_2O$) were prepared at the same concentrations as the cerium sulphate samples. For structural characterization, FT-IR measurements were carried out, in which the relative percentage of p-phase was estimated. For characterization of the electrical properties, dielectric constant measurements were carried out as a function of dopant concentration and frequency, in which different behaviours were observed for the samples doped with cerium in the two oxidation states (Ce^{3+} and Ce^{4+}). To characterize the optical properties, UV-Vis absorption and optical fluorescence spectroscopy measurements were carried out to verify the changes in these parameters caused by the addition of the dopant. The experimental results showed that the PVDF matrix doped with cerium sulphate and doped with cerium chloride are potential candidates for optical and photonic applications.

Chapter 1

1   Introduction

The search for multifunctional materials for technological applications has only intensified in the last 100 years since the quantum approach to matter. As a consequence, these increasingly sophisticated applications have demanded greater knowledge about the structure and properties of the materials used to produce the devices designed [1]. In the last century, a class of materials that has aroused great interest for technological applications has been polymeric materials, since they can be used in a wide variety of areas, replacing conventional raw materials such as metals, ceramics, glass, wood and fabrics. The synthesis of synthetic polymers has made it possible to obtain materials with a series of specific physical and chemical properties, which are often capable of equaling or even improving the performance of the material they replace. In this way, the use of polymers for technological applications has gained prominence, as they are easy to obtain, have low production costs and are lighter than conventional materials [2].

One group of polymers that has been intensively studied are the so-called ferroelectric polymers, including nylon, polyacrylonitrile (PAN), polyvinylidene cyanide (PVDCN), polyvinylidene fluoride (PVDF) and copolymers such as trifluoroethane (TrFE). Among the polymers with ferroelectric properties we can mention poly(vinylidene fluoride) or simply PVDF, which has been studied intensively since the 1960s because it manages to combine the characteristics of a plastic with those of a piezoelectric material. PVDF is a semi-crystalline material with four crystalline phases called a, P, γ and ô. Of these four crystalline phases, the most important is the P phase, as it is the phase that has ferroelectric properties and is widely used in the electro-electronic industry [3-9]. Despite being widely used in industrial applications, PVDF alone often does not meet the needs required for certain types of applications, making it necessary to develop composite materials that meet the growing demand from industry in general. In this sense, the so-called hybrid materials, formed from polymers and inorganic materials, have emerged as a solution to this problem, as they are designed to combine the desirable properties of two or more types of materials in order to present unique properties, They can exhibit improvements in stiffness, chemical and mechanical resistance, density, thermal stability, electrical and thermal conductivity and optical properties, such as transparency in the visible region of the electromagnetic spectrum [10,11].

In recent work, ferroelectric PVDF polymers have attracted great interest in the field of photonics for use as hosts for rare earth metal ions. The addition of rare earth elements improves the fluorescence properties so that the material can be used, for example, as an active medium for solid state lasers [12]. In this sense, various studies have been described showing the use of the so-called rare earth metals in materials such as glass, crystals and ceramics [13].

Therefore, the present work presents the synthesis and characterization of PVDF polymer doped with different concentrations of cerium in the Ce^{3+} and Ce^{4+} oxidation states in order to study the effects caused by the insertion of the dopant into the polymer matrix with a view to optical applications.

Chapter 2

2. Objectives

The aim of this work was to synthesize and characterize PVDF/Ce$_{(SO4)2.4H2O}$ and PVDF/CeCl3.7H2O composites for use in optical devices.

The characterization was done by measurements of:

- FT-IR to verify the incorporation of the dopant into the polymer matrix and the effects caused on the crystalline phases of PVDF;
- Dielectric constant for checking the electrical properties of composites as a function of frequency for different dopant concentrations in the polymer matrix;
- UV-Vis absorption and optical fluorescence spectroscopy measurements to verify the changes caused by the addition of dopants to the polymer matrix.

Chapter 3

3. Theoretical Background and Literature Review

3.1 Polymers

Polymers are macromolecules made up of repeating units called monomers and are obtained through chemical reactions called polymerization. The name polymer is derived from the Greek "poli" (*noXo*) meaning many and "meros" *(pépoq)* meaning parts or units that are repeated. These materials have a molecular weight that can vary from a few thousand to a few million atomic mass units. The polymer is represented by the structural unit placed in brackets or parentheses to represent the molecule (Figure 1), since it is impractical to try to represent the entire molecule [14-16].

$$\left[CH_2-CH_2\right] \qquad \left[CH_2-CH-CH_3\right] \qquad \left[CH_2-CH\right]$$

a   b   c

Figure 1 - Representation of the monomers of: (a) polyethylene; (b) polypropylene; (c) polystyrene [14-16].

The materials we now call polymers were first understood in 1920 when Staudinger, a German chemist, used the word 'macromolecule' to describe them. In the 1920s, Staudinger presented a paper that considered natural rubber and other products made up of long chains rather than small colloidal aggregates to be macromolecules. At first, this concept of macromolecule was not well received by the scientific community, but in 1928 it was recognized that polymers were substances of high molecular weight [16].

Polymeric materials can be divided into two large groups: the first are called natural polymers and the second are synthetic polymers. Natural polymers were the materials used as standards by scientists for the synthesis of synthetic polymers. The synthesis of synthetic polymers began mainly after World War II, more specifically at the beginning of the 1950s, because in this period (post-war) there was a growing need to synthesize new materials for technological applications [17].

In 1935, American chemist Wallace Carothers invented naylon at the Dupont company while carrying out basic research into polymerization reactions. From this research, Carothers showed the potential of synthetic polymers, a potential that would become a reality in an extraordinarily short time. In the 1950s, polymer synthesis was completely mastered by Karl Ziegler and Giulio Natta, through the discovery of catalysts based on alkylmetallic systems, used in the synthesis of modern plastics with wide industrial application. This discovery was so important that today synthetic polymers are used in greater quantities than any other class of material [18]. Examples of this use include: thermally stable polymers used in aerospace applications; engineering plastics; non-flammable polymers; degradable polymers; polymers for medical applications and electrically conductive polymers, among others [14].

The dominant influence on the behavior of a polymer is its morphology, which depends on chemical aspects, but the physical properties that can be analyzed are those resulting from the arrangement of the polymer chains and their response to deformation forces. The arrangement of the polymer chain in a unit cell can be verified using the X-ray diffraction technique. The thermal stability of a polymeric material is verified by two main thermal transitions: the first is called the glass transition temperature, T_g, and the second is called the crystalline melting temperature, T_m. The glass transition temperature is defined as the temperature at which the amorphous domains of a polymer acquire mobility and the melting temperature of the crystalline domains of a polymer. The presence of these two transitions or just one of them depends on the morphology of the polymer sample. If the polymer is completely amorphous it will only show T_g, if it is completely crystalline it will only show T_m, and if the polymer is semi-crystalline it will show both thermal transitions [16, 19, 20].

3.1.1 Classification of Polymers

Polymers can be classified in different ways. In terms of origin, a polymer can be classified as natural or synthetic. With regard to the variety of monomers used in their formation, they can be classified as homopolymers or copolymers [1, 2, 16]. Homopolymers are polymers formed from a single monomer. Homopolymers are usually synthesized through bonding reactions. Examples include polyethylene (PE), polypropylene (PP) and polyvinylidene fluoride (PVDF) (Figure 2). Copolymers, on the other hand, are formed from two or more different monomers, which are called comonomers. Depending on the arrangement in which the monomers appear in the chains, copolymers can be divided into different groups and can form random, alternating, block or grafted copolymers [16].

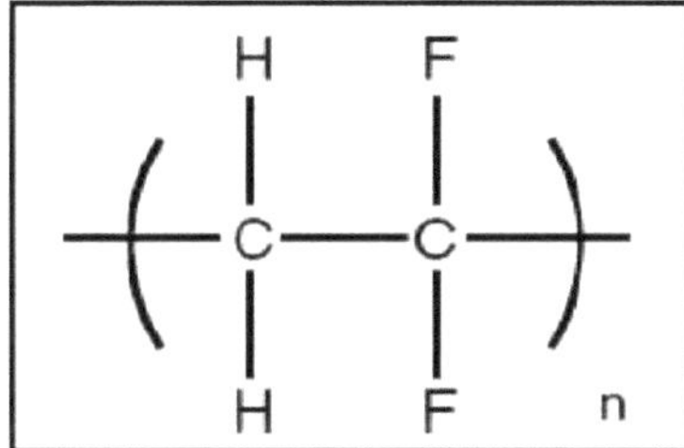

Figure 2 - Repetitive unit of PVDF [16].

a) Random, random or statistical copolymers.

These are polymers in which the meres do not follow a defined order in the polymer chain, i.e. the monomers bind randomly during polymerization. Since C and D are different monomers, we have an example of a schematic representation of part of their structure:

~~~~**C-C-D-D-D-D-D-C-C-D-C-D-D-C**~~~~

b) Alternating copolymer.

They are polymers in which the meres appear alternately in the polymer chain and are generally formed by condensation polymerization reactions, such as the reaction between a diacid and a dialcohol or a diamine:

~~~~**C-D-C-D-C-D-C-D-C-D-C-D-C-D-C-D**~~~~

c) Block copolymer.

They are polymers that have long sequences of one of the moieties linking with long sequences of the other, so that each one appears in blocks of homopolymers linked together:

$$\sim\sim C\text{-}C\text{-}C\text{-}C\sim\sim D\text{-}D\text{-}D\text{-}D\sim\sim C\text{-}C\text{-}C\text{-}C\sim\sim D\text{-}D\text{-}D\text{-}D\sim\sim C\text{-}C\text{-}C\text{-}C\sim\sim D\text{-}D\text{-}D\text{-}D\sim\sim$$

d) Grafted copolymer.

These are polymers that have a homopolymer main chain (poly C) with another covalently linked homopolymer chain formed by another monomer (poly D):

$$-(C\!-\!C\!-\!C\!-\!C\!-\!C\!-\!C\!-\!C\!-\!C\!-\!C\!-)_n$$
$$|$$
$$(D\!-\!D\!-\!D\!-\!D\!-\!D\!-)_m$$

Polymers also have some other classifications. One of these uses the response of polymers to heat treatment, dividing them into thermoplastics and thermosets. Thermoplastics are polymers that melt when heated and solidify again when cooled. Thermo-rigid polymers are those that do not melt when heated, but at sufficiently high temperatures undergo decomposition [17].

Polymers can be classified based on the nature of the chemical reactions used in polymerization. This method was suggested by Carothers in 1929. Based on this classification, the two main groups are adhesion polymers and condensation polymers [17]. Adhesion polymers always have the same atoms as the monomers in their repeating units, while condensation polymers have fewer atoms in their repeating units than the original monomers, due to the formation of by-products during the polymerization process. Classifying a polymer as an adhesion or condensation polymer may be inappropriate, as some polymers can be obtained from different reagents, and consequently, according to Carothers' classification, they could receive different classifications depending on the synthetic route used [14]. Examples are:

a)   Forming polyether from ethylene oxide or ethylene glycol (Figure 3):

a) $H_2C\overset{O}{\overgroup{-\!\!-}}CH_2 \longrightarrow -\!\!\left[CH_2CH_2O\right]\!\!-$

b) $HOCH_2CH_2OH \longrightarrow -\!\!\left[CH_2CH_2O\right]\!\!- + H_2O$

Figure 3 - Formation of polyethers: (a) by adhesion; (b) by condensation [14].

A polyester from a lactone or ro-hydroxycarboxylic acid (Figure 4):

a) $\underset{R}{O\!-\!CO} \longrightarrow -\!\!\left[O\!-\!R\!-\!CO\right]\!\!-$

b) $HO\!-\!R\!-\!CO_2H \longrightarrow -\!\!\left[O\!-\!R\!-\!CO\right]\!\!- + H_2O$

Figure 4 - Forming of polyesters: (a) by adhesion; (b) by condensation [14].

c) A polyamide from a lactone or ro-aminocarboxylic acid (Figure 5):

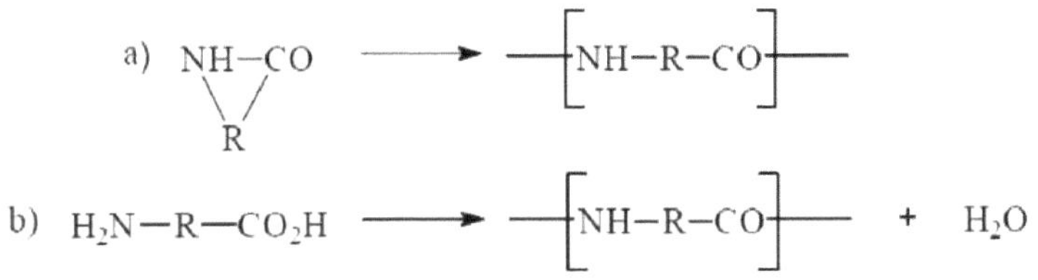

a) NH—CO $\longrightarrow$ —[NH—R—CO]—
     \ /
      R

b) H_2N—R—CO_2H $\longrightarrow$ —[NH—R—CO]— + H_2O

Figure 5 - Forming of polyamides: (a) by bonding; (b) by condensation [14].

d) A polyurethane from a diisocyanate and an alcohol or from a diamine and a dichloroformate (Figure 6):

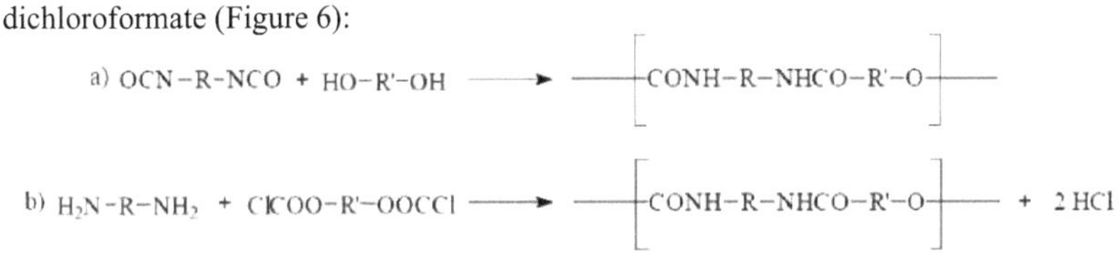

a) OCN—R—NCO + HO—R'—OH $\longrightarrow$ —[CONH—R—NHCO—R'—O]—

b) H_2N—R—NH_2 + ClCOO—R'—OOCCl $\longrightarrow$ —[CONH—R—NHCO—R'—O]— + 2 HCl

Figure 6 - Polyurethane formation: (a) by adhesion; (b) by condensation [14].

Today, to avoid this ambiguity, polymerization reactions are commonly classified according to the polymerization mechanism. If the polymer chains are formed step by step by joining monomer molecules to form dimers, trimers, and larger species from the parent monomer, the process is called step polymerization or step-growth polymerization. If, on the other hand, the molecular weight increases by successive attachments of monomer molecules to the end of a growing chain, the process is called chain-reaction or chain-growth polymerization.

3.2 Ferroelectric materials

Ferroelectricity was first observed in a monocrystalline material known as Rochelle salt (double sodium potassium tartrate) in 1921 by Joseph Valasek. Since the discovery of ferroelectricity in a polycrystalline material (barium titanate ceramic) in 1944, significant progress has been made in understanding the properties of this class of materials, as well as possibilities for technological applications. Ferroelectric materials are materials in which the center of the positive charges does not coincide with the center of the negative charges, resulting in the formation of intrinsic dipoles. The application of an external electric field is capable of causing the dipoles to twist, thus producing a state of polarization in the sample even after the field has been removed. It is important to note that individual dipole rotations do not normally occur, but rather groups of dipoles with resulting dipole moments that undergo this rotation. These groups are called domains [21-23]. Figure 7 shows a characteristic graph of polarization by applied electric field. This graph is known as the hysteresis lake and shows exactly that in ferroelectric materials there is a remaining polarization, something that in a normal crystal would not be observed.

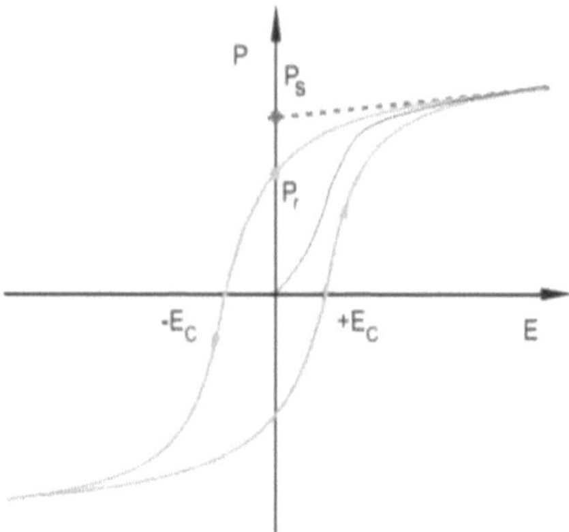

Figure 7 - Representation of the polarization hysteresis (P) of a typical ferroelectric material as a function of the applied electric field (E) [24].

In ferroelectric materials, the application of high electric fields leads to the so-called saturation polarization (P_S) and if the applied field is removed, the polarization does not return to zero, defining the so-called remaining polarization (P_r). The value of the applied electric field whose spontaneous polarization is equal to zero is called the coercive field (E_c). In these materials there are temperatures at which structural phase transformations occur. In ferroelectric materials, the threshold temperature between the ferroelectric state (polar) and the paraelectric state (apolar) is called the Curie temperature. The ferroelectric phase occurs below the Curie temperature, and the paraelectric phase occurs above the Curie temperature. In the paraelectric phase there is no ferroelectric effect for these materials, but it can be achieved by applying an electric field of the order of KV/m. Piezoelectric materials are materials that become electrically polarized when subjected to a mechanical stress, or by undergoing mechanical deformation when subjected to the action of an electric field. The phenomenon of pyroelectricity consists of a change in electrical polarization resulting from changes in temperature [25,26]. The major difference between the two effects is that piezo consists of a primary effect, while pyroelectricity can be a secondary effect to the first. To put it simply, while piezo depends on direct mechanical deformation, pyroelectricity can be the result of the expansion or contraction of the material (mechanical effect) resulting from the temperature difference which causes a change in potential or vice versa. Pyroelectricity can also appear without expansion or contraction of the material, as a variation in the entropy or internal configuration of the material.

The first studies on piezoelectric properties in polymers took place in the 1950s, on polymers of biological origin, such as collagen and cellulose, and optically active synthetic polymers, such as polyamides and polylactic acids. In 1969, Kawai showed that the polymer poly(vinylidene fluoride), after being stretched and polarized in high electric fields, has higher piezoelectric coefficients than quartz. Two years later, Bergman and colleagues observed pyroelectric activity in polyvinylidene fluoride films. In the same decade, PVDF was shown to have ferroelectric hysteresis curves, but it was only in 1980 that PVDF was accepted as a ferroelectric material, after a detailed study of its ferroelectricity [21].

3.2.1 Polyvinylidene Fluoride

Polyvinylidene fluoride, PVDF, is a thermoplastic polymer, which at room temperature is in the solid state with a melting temperature (*Tm*) in the range of 165 to 179°C and a glass transition temperature (*Tg*) of around -34°C. The relatively low melting temperature range and some of the properties already described ensure that the polymer can be easily processed by melting and mechanical mixing, which is a great advantage in terms of large-scale production. Its high stability to chemical and mechanical agents means that PVDF is widely used in packaging for corrosive chemical products and in valve diaphragms [27]. In 1969, Kawai's work projected PVDF as an electrically active material, reporting extraordinary piezoelectric properties [3]. The concept of a pyroelectric material was also added with the work of Bergmam et al (1971) and Nakamura and Wada (1971) [28,29]. The high pyroelectric and piezoelectric properties of PVDF, combined with its malleability, impact resistance, flexibility and low mechanical impedance made it possible to launch, on an industrial scale, the first polymeric electromechanical transducer, which still prevails in the market today, with recurrent use from headphones to ultrasound sensors in submarines, among other civilian and military applications [30,31].

PVDF is produced by the polymerization of VDF. The monomer was first synthesized at the beginning of the last century (Figure 8), is a gas under normal temperature and pressure conditions and can form explosive mixtures with air. It is used both to produce PVDF and as a comonomer, together with other fluorinated monomers such as hexafluoropropylene (HFP), chlorotrifluoroethylene (CTFE) or tetrafluoroethylene (TFE) to produce various fluorinated elastomers or copolymers [32,33].

$$HC \equiv CH + 2HF \xrightarrow{BF_3} CH_3CHF_2$$

$$CH_3CHF_2 + Cl_2 \longrightarrow CH_3CClF_2 + HCl$$

$$CH_3CClF_2 \longrightarrow H_2C = CF_2 + HCl$$

Figure 8 - Production steps for VDF monomer [32].

The repeating unit of PVDF is formed by the polar structure $-CH_2-CF_2$. Conventionally, the electropositive end - CH_2 - is called the *head* and the electronegative end - CF_2 - is called the *tail*. When PVDF is polarized, there is a tendency (of an electrical nature) for *head-tail* type bonds to form between the moieties that form linear chains. The formation of irregular bonds between the *head-to-head* or *tail-to-tail* repeating units are considered structural defects. Structural defects are chemical in nature and are permanent throughout the life of the polymer chain. The formation of these defects causes a decrease in the local dipolar moment of the chain, and thus frustrated dipolar interactions arise, affecting the structural stability of the crystalline phase [34].

The electric dipole originating from CF_2, locally perpendicular to the carbon chain, is significantly responsible for the behavior of the electrical interactions between PVDF chains. The process of minimizing electrostatic energy (interaction energy between dipoles) leads to a process of local pairing between the chains. This pairing can establish a periodicity in the polymeric material that is much larger than the size of the repeating unit. These regions of finite periodicity are actually small molecular crystals, called crystallites. The crystallites are distributed throughout the material and interconnected by amorphous material, defining a semi-crystalline material. The fraction of the material occupied by the crystallites is known as the sample's degree of crystallinity. The degree of crystallinity of PVDF typically ranges from 50% to 68% and is susceptible to external factors such as temperature or pressure [27, 35].

Figure 9 shows a structure called a spherulite, a radial formation formed by PVDF, which occurs more frequently in samples crystallized slowly at temperatures close to melting. The branches that form the spherulites are called fibrillar branches that contain the lamellae, which in turn contain the crystallites. The occurrence of crystallites is independent of the recrystallization temperature, even in the absence of spherulites. The quality of the crystallites is strongly linked to the amorphous matrix.

Dipolar interactions within and between chains determine the formation of crystallites. The formation of crystallites minimizes the electrostatic energy of the system, but increases the elastic potential energy of the amorphous phase as a result of increased entanglement and tension on the chain segments that interconnect the crystallites, thus defining an intrinsic delimiter of the degree of crystallinity in linear polymers [36]. Recent developments in the Langmuir-Blodgett technique have made it possible to grow self-constructed polymer films, overcoming this limitation [37, 38].

      The pyroelectric and piezoelectric properties depend on the structure of the PVDF, since the crystal structure depends on the conformation of the polymer chains (Figure 10). These are regular sequences of angles of 180° (trans bonds) and ± 60° (gauche+ and gauche$^-$ bonds). The trans and gauche angles decrease the dipolar and volumetric interaction energy [39,40].

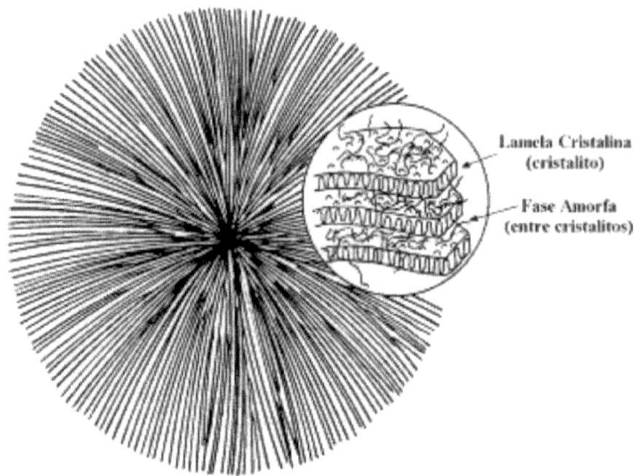

Figure 9 - Representation of the structures presented by PVDF: spherulites, lamellae, crystallites and amorphous matrix [39].

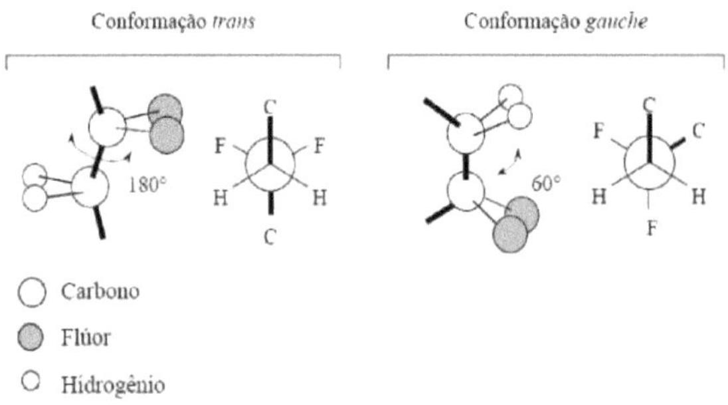

Figure 10 - Trans and gauche angular conformations of a linear polymer. They are shown in profile and parallel to the carbon-carbon bond (Newman projection) [39].

      The angular combinations responsible for the formation of crystalline structures in the PVDF (Figure 11) are all-trans (or ttt), tg$^+$ tg$^-$ and t3g$^+$ t3g$^-$ (or tttg$^+$ tttg$^-$).

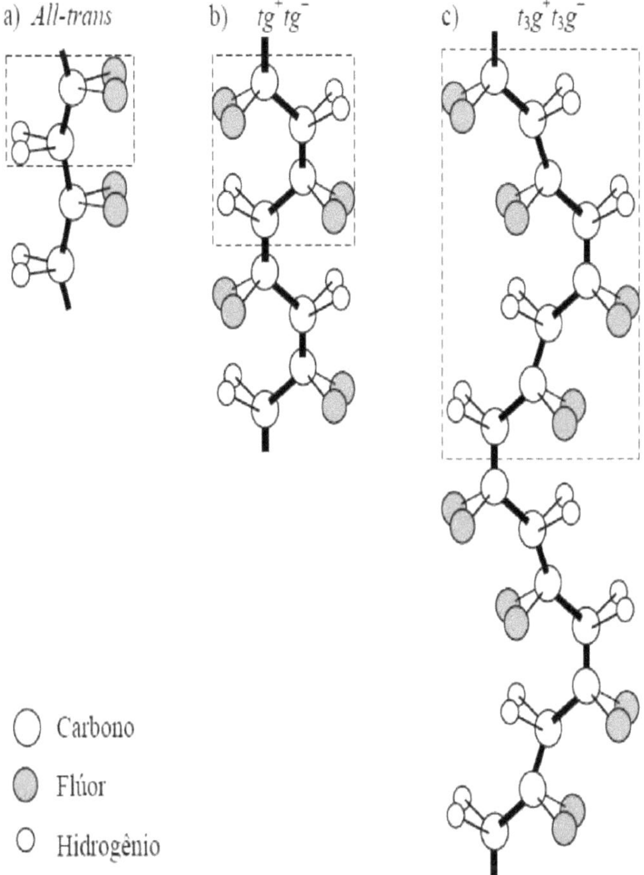

Figure 11- PVDF chains in angular conformations: (a) ttt (alltrans), (b) tg+ tg and (c) t3 g+ t g3⁻ . The respective repeating units are highlighted [39].

Among synthetic polymers, PVDF is unique in that it has a complex crystalline polymorph, with at least five distinct crystalline phases, called alpha, beta, gamma, delta and epsilon phases. The different phases are formed according to the polymer's processing method. The most common phases are the alpha and beta phases, formed during conventional processing, while the gamma and delta phases are obtained under special conditions [32].

3.2.2 Crystalline phases of PVDF

The **alpha** phase is characterized by its orthorhombic crystal structure and apolar characteristics. It is formed by moderate or rapid cooling of the melt, and can also be obtained by evaporation of the solvent. The chains are arranged in a helical trans-Gauchy (TGTG') configuration, which allows for greater distance between the fluorine atoms. The alpha phase is apolar due to the packing of the chains giving rise to opposite parallel dipole moments, as shown in Figure 12 [34,41,42].

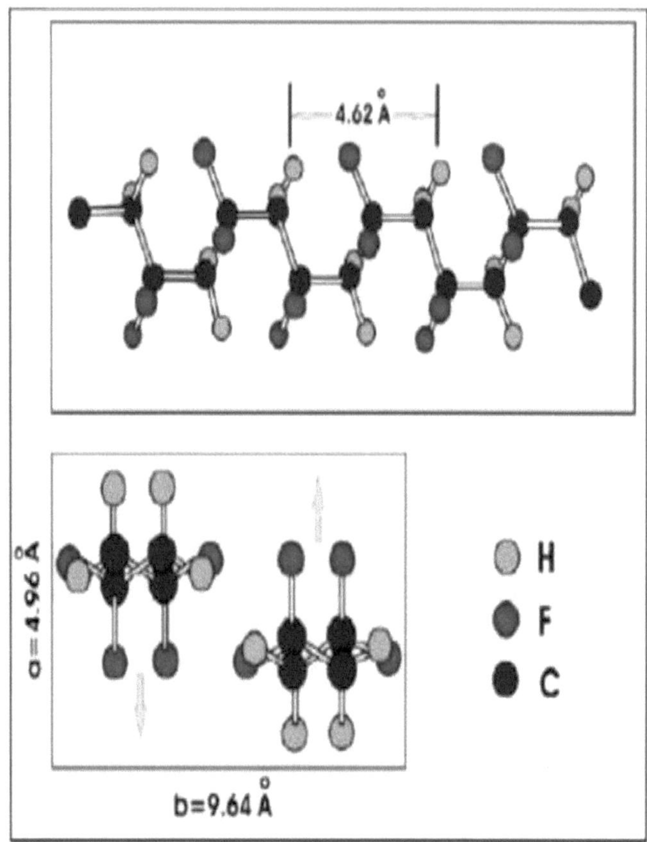

Figure 12 - Representation of the structure and unit cell of PVDF in the alpha phase [41,42].

The **beta phase** is characterized by planar zigzag bonds, formed by trans-trans (TT') bonds in the main chain, normally produced from the controlled stretching of the alpha phase. It is a polar phase, with an orthorhombic unit cell, in which the fluorine atoms are arranged opposite the hydrogen atoms, as shown in Figure 13, with a slight deviation caused by the deflections of adjacent carbon-fluorine bonds [34,41].

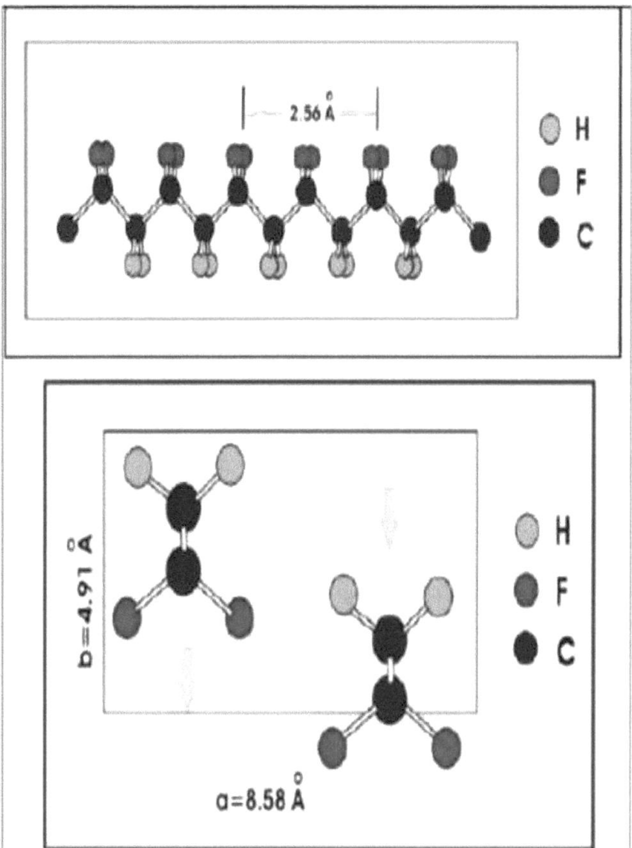

Figure 13- Representation of the structure and unit cell of PVDF in the beta phase [41,42].

This phase has strong piezoelectric and pyroelectric properties due to the dipole formed by the orientation of the fluorine atoms, making it widely studied for applications in technology, in the form of sensors, transducers and polymer actuators [41].

The **gamma phase** can be obtained by crystallizing the polymer in solution or in the melt at temperatures above 155°C for periods of more than 6 hours. This phase is characterized by the presence of gauche bonds, interspersed with three trans bonds (TTTGTTTG') leading to the formation of a unit cell with a monoclinic structure with polar characteristics, as shown in Figure 14 (with hydrogens omitted) [41].

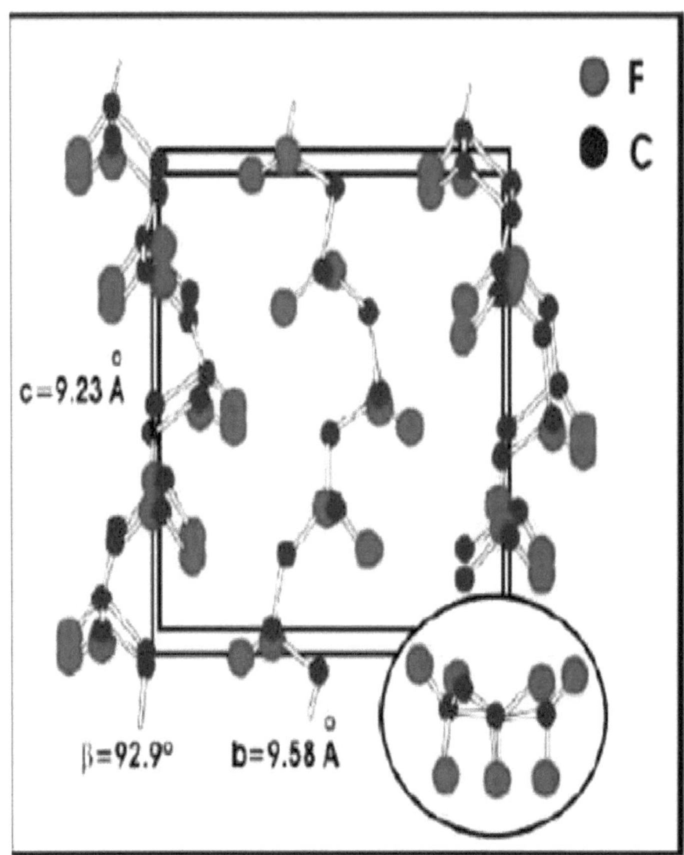

Figure 14- Representation of the structure and unit cell of PVDF in the gamma phase [41,42].

The **delta phase** can be obtained by exposing the alpha phase to an intense electric field, which induces the inversion of the electric dipoles of the chains organized in the alpha phase, forming a polarized version of it. This gives it the same conformational structure as the alpha phase (TGTG'), with differences in the way the chains are packed, forming an orthorhombic unit cell, as shown in Figure 15 [34, 41,43]

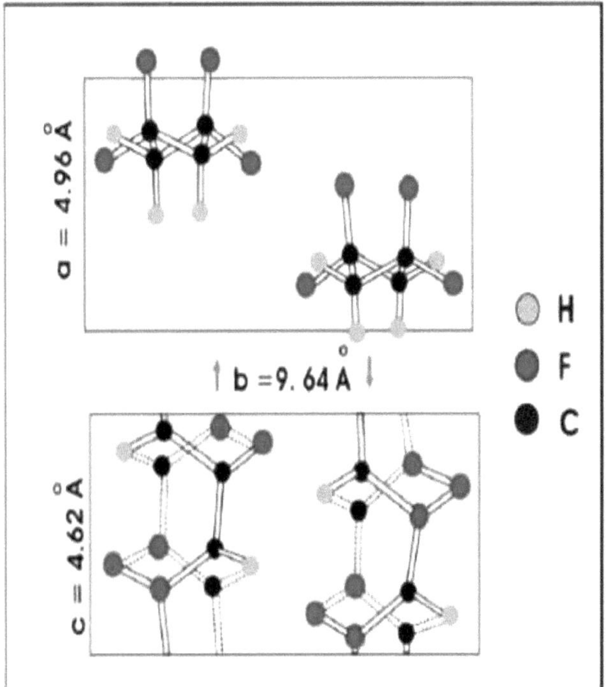

Figure 15- Representation of the structure and unit cell of PVDF in the delta phase [41,42].

The **epsilon phase** is the least common phase in PVDF. It has a t3g⁺ t3g⁻ arrangement, similar to the gamma phase, but arranged antiparallelly, resulting in an apolar structure. This phase would only be stable just below the melting temperature. The literature often describes this phase as gamma-polar PVDF [44].

3.3 Cerium

The ions formed by rare earth metals are generally in the three oxidation state, which gives rise to relatively large ions with radii of approximately 1Å. Coordination numbers range from 6 to 12, with the predominant number being eight [45]. Rare earths have a wide range of applications, being used as catalysts in the treatment of automobile emissions and in oil cracking, in the manufacture of lasers, as luminescent materials in the manufacture of light bulbs and cathode ray tubes, in metallurgy, magnets, in the ceramics and glass industries and in biological systems [45,46]. Among the rare earths we can mention Cerium, a chemical element with atomic number 58, one of the 15 elements classified as lanthanides (La to Lu), which together with Sc and Y form the rare earth family. Cerium is the most abundant of the rare earths. Its abundance in the Earth's crust is around 60 ppm, making it the 26th most abundant element in terms of frequency of occurrence, almost as abundant as copper and nickel. Despite being the most abundant lanthanide, cerium is found in many minerals only at trace levels, with bastinasite and monasite being the most important minerals as a source of cerium and supplying most of the world's demand for this element [47-52]. Cerium has the electronic configuration [Xe]4f 5d s^{112} , is very electropositive and its interactions are predominantly ionic as it has a low ionization potential (3.49 kJ mol^{-1}) for the removal of the first three electrons [47,49]. Like the other lanthanides, the most stable oxidation state of this element is (III). The properties of trivalent cerium are similar to other lanthanide ions with the same oxidation state [49,51], with the exception of its easy oxidation to Ce^{4+} and instability in air and water. Its salts, which are generally more stable than those of Ce^{4+} , are not very hydrolyzable and are widely used as precursors for various cerium compounds [47-52]. Unlike the other elements in this class, the (IV) oxidation state is also stable, mainly due to its electron configuration similar to that of a noble gas ($[Xe]4f0$). This element is the only lanthanide in the tetravalent state that is stable in aqueous solution, but the influence of the larger charge and smaller ionic size means that the salts of the Ce ion^{4+} are more easily hydrolyzed in aqueous solutions than the Ce ion^{3+} and, as a result, these solutions are strongly acidic [51]. In the tetravalent state, cerium is a strong oxidizing agent and can be reduced, for example, by organic acids, ferrous salts, hydrogen peroxides and other inorganic and organic compounds [51], since the standard reduction potential (E°) for the Ce(IV)/Ce(III) pair is approximately 1.6 V (with some variations depending on the anion and the medium) [51]. Despite this potential, aqueous solutions containing Ce(IV) are stable, probably due to kinetic effects. Because of these oxide-reduction properties, cerium is well known and used in a number of organic synthesis reactions [51,53] and analytical methods [51].

Among the Ce compounds^{4+} , cerium dioxide (CeO_2, also called ceria) is the most stable due to its fluorite-type cubic structure (space group Fm3m) [49,51]. This structure gives this oxide greater stability than the sesquioxide, Ce_2O_3, with trivalent cerium (hexagonal structure) [49]. CeO_2 can be obtained from precursor salts by calcination in air or an oxygen-containing medium; non-stoichiometric cerium oxide, CeO_{2-x} (X can be above 0.3), can also be obtained, in which case dopants are used, such as oxides of other rare earths like Y or La. The introduction of these elements into the cerium dioxide crystal lattice causes charge compensation to occur, generating oxygen vacancies, and the mobility of the oxygen depends on the type of dopant used [47-52].

Cerium compounds have enormous potential for application in various technological fields, mainly due to some of their properties, such as redox potential range, high oxygen mobility in the crystal lattice, high affinity for compounds containing oxygen, nitrogen and sulphur [47,54]. Some of these applications are shown in the flowchart in Figure 16.

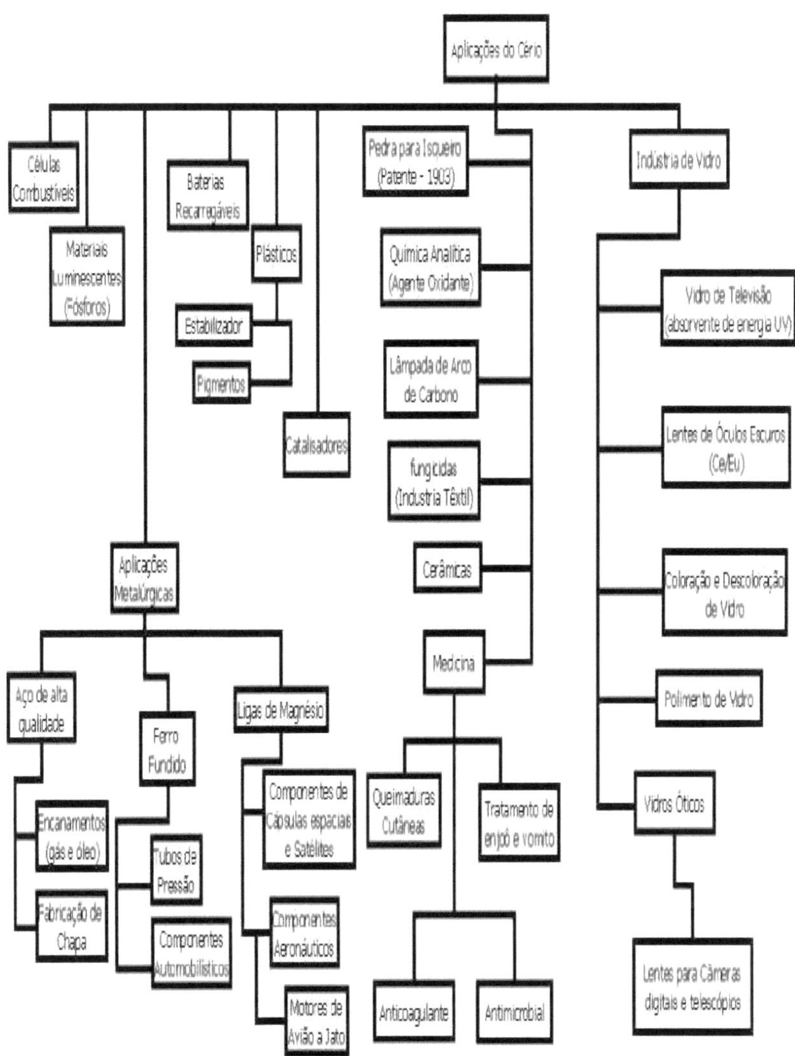

Figure 16- Flowchart of some applications of cerium compounds [47,54].

One of the main applications of cerium-based compounds is in metallurgical processes, where these compounds are added to auger to eliminate impurities, mainly oxygen and sulphur, due to their high affinity for forming bonds with these elements, thus helping to significantly improve the quality of the auger, for example by increasing strength and ductility [50,54]. For the same purpose, these compounds can also be added to other metal alloys, such as cobalt or nickel-based alloys, which are used, for example, in aircraft engine turbines [55,56]. Cerium, in the form of salts, chlorides and nitrates, is also used to prevent corrosion of aluminum utensils (and their alloys), replacing chromate ions, which have negative implications from an environmental point of view [51,54]. Another important application of cerium is in fuel cells. These cells are of great importance today, especially in the search for alternatives for generating electricity in a clean and efficient way. A general diagram of how a fuel cell works is shown in Figure 17. Cells with a solid oxide electrolyte are based on a conductive electrolyte of oxygen ions, through which the O ions^{2-} migrate from the cathode to the anode, where they react with a fuel, generating an electric current in the system. In the case of cerium, this ionic conduction is due to the effects of the oxygen vacancies in the crystal structure of CeO2. There are several other types of O^{2-} ion conductors, the most traditional of which are zirconia-based, but CeO_2 electrolytes have been preferred due to their high ionic conductivity, good compatibility with other cell constituent materials and ability to operate in a lower temperature range (between 500 and 700°C, compared to around 1000°C required when using traditional electrolytes) [57,58].

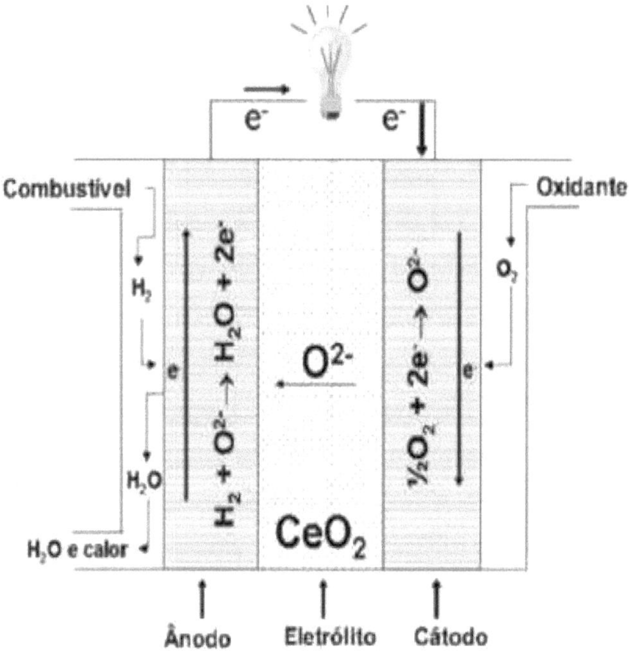

Figure 17 - Schematic representation of how a fuel cell works

3.4 Composites

Many of our modern technologies require materials with unusual combinations of properties that cannot be met by conventional metal alloys, ceramics and polymeric materials, making it necessary to develop composite materials. In general, a composite can be considered to be any multiphase material that exhibits a significant proportion of the properties of its constituent phases, in such a way that a better combination of properties is obtained. In other words, a composite is any material made up of two or more different constituents, which offers properties that cannot be obtained with the pure components.

The vast majority of composite materials are made up of just two phases; one is called the matrix, which is continuous and surrounds the other phase, often called the dispersed phase. The properties of composites depend on the properties of the constituent phases, their relative quantities and the geometry of the dispersed phase. In this context, "geometry of the dispersed phase" means the shape of the particles, their size, distribution and orientation [59]. The incorporation of inorganic particles into polymers gives rise to materials with greater mechanical strength, thermal stability or optical, magnetic and electrical properties superior to those of the pure components [60-62]. Particles with nanometric dimensions have a high surface area, promoting better dispersion in the polymer matrix and consequently an improvement in the physical properties of the composite, which depend on the homogeneity of the material.

Organic-inorganic hybrid materials are a special class of multifunctional materials that have received a lot of attention in recent years [63-67]. The specific architecture of these materials provides a synergistic effect between the organic and inorganic counterparts, giving rise to compounds with physical and/or chemical properties distinct from their isolated components. These compounds not only represent a creative alternative for researching new materials, but also allow for the development of innovative industrial applications. Potential applications for organic-inorganic hybrid materials, i.e. for composites combining ceramics and polymers, include: intelligent membranes, photovoltaic devices, fuel cells, photocatalysis, biosensors, intelligent microelectronic devices, microoptics, new cosmetics, controlled release of drugs, controlled release of active molecules, etc.

Compared to unmodified polymers, hybrid materials formed from polymers and oxides can exhibit improvements in their properties, such as stiffness, chemical and mechanical resistance, density, impermeability to gases, thermal stability, electrical and thermal conductivity, as well as a high degree of optical transparency [63]. The first successful development of lamellar inorganic solids and polymers was carried out by Toyota researchers for structural applications in vehicles. They prepared composites from the combination of nylon 6 and clay using the *insitu* polymerization method.

Chapter 4

4 Materials and Methods

To prepare the PVDF films, a study was carried out on the best conditions using the solution crystallization method, in order to obtain films with a higher percentage of beta phase (a phase desirable for its ferroelectric properties).

To obtain the films, a solution of PVDF (supplied by Solef 11010/1001) with a concentration of 75 mg/mL was prepared. The solution was prepared by dissolving the polymer in dimethylformamide (DMF - supplied by Sigma Aldrich, 99.8% purity) using magnetic stirring for 60 minutes at room temperature. After preparing the solution, aliquots containing 150 mg of the polymer were transferred to test tubes and shaken for 1 minute for better homogenization. The aliquots were then transferred to Petri dishes and subjected to heat treatment (50° C) in a drying oven for 16 hours to remove the solvent and form the films.

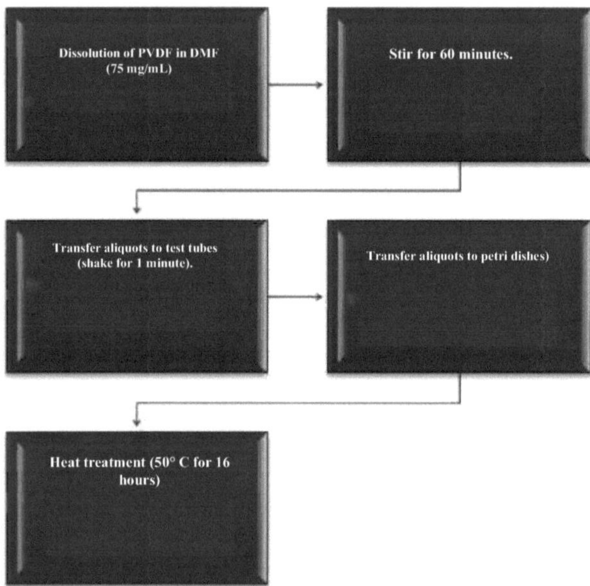

Figure 18 - Schematic diagram of the preparation of PVDF films.

4.2 Synthesis of the composites

The second stage of the work was the synthesis of PVDF/CeSO4.4H2O and PVDF/CeCl3.7H2O composites (Figures 19 and 20). For the synthesis of the composites, solutions of PVDF were prepared at a concentration of 75 mg per milliliter of DMF, solutions of cerium (IV) sulfate tetrahydrate with a purity level of 99. 8% (supplied by Sigma Aldrich),8% (supplied by Sigma Aldrich) at a concentration of 0.6 mg per milliliter of DMF and solutions of cerium (III) chloride heptahydrate at a purity of 99.9% (supplied by Sigma Aldrich) at a concentration of 0.6 mg per milliliter of DMF.

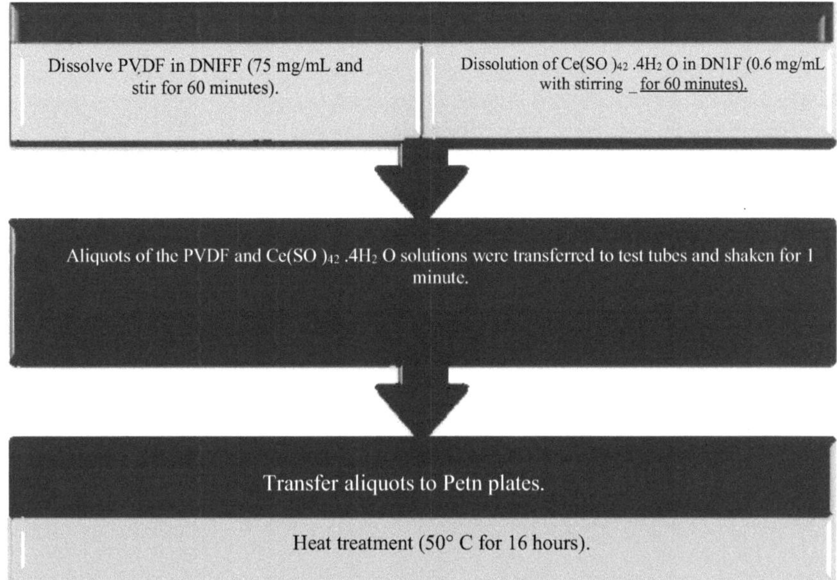

Figure 19- Schematic diagram of the preparation of PVDF/CeSO4.4H2O composites.

The solutions were prepared using magnetic stirring for 60 minutes at room temperature. After preparing the solutions, an aliquot containing 150 mg of PVDF was pipetted into a test tube and shaken for 1 minute for better homogenization. The aliquot was then transferred to a Petri dish (blank sample). The PVDF/CeSO4.4H2O composites were prepared by pipetting aliquots of the polymer solution and aliquots of the solution of cerium (IV) sulphate tetrahydrate in percentages of 0.2, 0.4, 0.6, 0.8 and 1.0% by mass respectively and transferring them to test tubes and shaking for 1 minute for better homogenization. After shaking, the aliquots were transferred to Petri dishes. The PVDF/CeCl3.7H2O composites were prepared by pipetting aliquots of the polymer solution and aliquots of the solution of cerium (III) chloride heptahydrate in percentages of 0.2, 0.4, 0.6, 0.8 and 1.0% by mass respectively and transferring them to test tubes and shaking for 1 minute for better homogenization. After shaking, the aliquots were transferred to Petri dishes. After preparation, the blank sample and the $PVDF/CeSO_4 .4H_2 O$, PVDF/CeCl3.7H2O solutions were taken for heat treatment (50° C) in a drying oven for 16 hours.

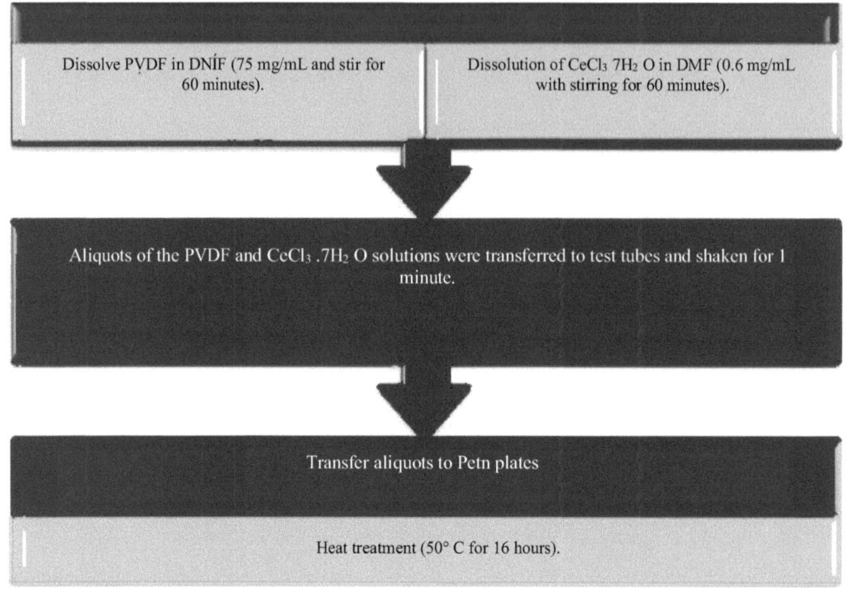

Figure 20 - Schematic diagram of the preparation of PVDF/CeCl3.$7H_2$ O composites.

4.3 Characterization by Transform Infrared Spectroscopy of Fourier

The measurements were carried out on an FT-IR - 4100 spectrophotometer, Jasco Corporation, at room temperature, with a resolution of 2 cm^{-1} , 250 scans per measurement and a spectral range of 4000 to 500 cm^{-1} . The materials were analyzed directly because they were already in film form. The CeSO4.4H2O and CeCl3.7H2O compounds were analyzed in the form of KBr pellets.

Figure 21- Spectrophotometer used for FT-IR measurements of the composites PVDF/Ce(SO4)2.4H2O and PVDF/CeCl3.7H2O.

4.4 Characterized by UV-Vis Spectroscopy

The absorption spectra of the composites were taken on a Varian Cary 50 UV-Vis spectrophotometer. The absorption measurements were made in the wavelength range (!) from 200 to 800 nm.

Figure 22 - Spectrophotometer used for UV-Vis measurements of the composites PVDF/Ce(so4)2.4H2O and PVDF/CeCl3.7H2O.

4.5 Characterized by Fluorescence Spectroscopy

Fluorescence spectroscopy was used to evaluate the influence of dopants on the emission spectrum of the polymer matrix. The fluorescence measurements were carried out on a portable spectrofluorimeter consisting of lasers operating at 405 nm and 532 nm, a monochromator (USB 2000 FL/ Ocean Optics), a Y-type fiber and a laptop computer. For the fluorescence measurements carried out in this study, only the 405 nm wavelength excitation laser was used.

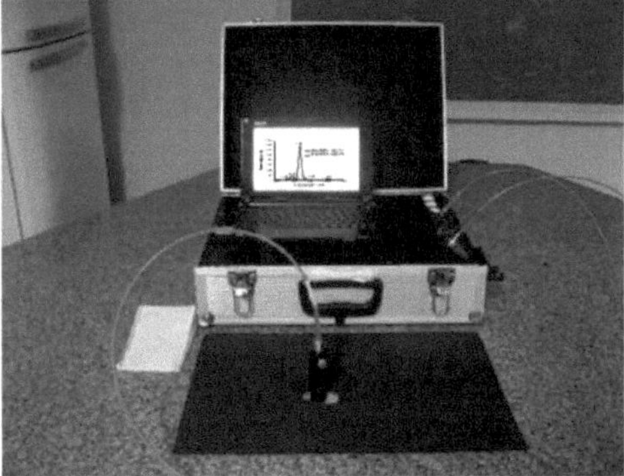

Figure 23- Portable Fluorimeter used to measure the fluorescence of the composites PVDF/Ce(so4)2.4H2O and PVDF/CeCl3.7H2O.

4.6 Dielectric Characterization

An Agilent E4980 RLC bridge was used for the samples studied in this work. The dielectric constant measurements were carried out in the background at the laboratory of the Multifunctional Device Development Group of the Physics Department of the State University of Maringá.

Figure 24 - Agilent E4980 RLC bridge used for dielectric constant measurements of PVDF/Ce(SO4)2.4H2O and PVDF/CeCl3.7H2O composites.

Chapter 5

5. Results and Discussions

5.1 Infrared Spectroscopy - PVDF/Ce(so4)2.4H2O

Table 1 shows the main characteristic bands of the amorphous phase and the crystalline phases of PVDF $(\alpha,\ \beta\ e\ \gamma),$ with their respective phase vibration modes. crystalline [68-75].

Table 1 - Characteristic Vibrational Modes of PVDF.

Band (cm)$^{-1}$	Phase	Vibrational Group and Mode
511	β	Deformation (CF_2) [75]
512	γ	Deformation (CH_2) [72]
530	α	Deformation (CF_2) [75]
600	Amorphous	[73]
615	α	Deformation (cf2) and deformation in the Skeleton [75]
678	-	Defects in the polymer chain due to bonds cabega-cabega or cauda-cauda [75]
766	α	Deformation (cf2) and deformation in the Skeleton [75]

778	γ	Balango (CH_2) [75]
795	α	Balango (CH_2) [75]
812	β ou γ	Deformation (CH_2) out of plane [75]
833	γ	
840	β ou γ	Balango (CH_2) [75]
855	α	Out-of-plane deformation (CH) [75]
877	β	Balango (CH_2) [75]
880	Amorphous	[73]
976	γ	Out-of-plane deformation (CH) [75]

The infrared spectra obtained for the PVDF films as a function of dopant concentration are shown in Figure 25. These measurements were taken to verify the incorporation of the dopant into the polymer matrix and to analyze the effects caused by them from a structural point of view.

The FT-IR spectra (Figure 25) show the characteristic peaks of the beta phase at (840, 812 and 511 cm^{-1}) and the alpha phase (976, 855, 796, 764, 614 and 532 cm^{-1}) [72,73]. The characteristic bands of the amorphous portion of the polymer at 600 and 880 cm^{-1} can also be observed in all the samples [76,77]. With the addition of cerium sulphate, significant changes can be observed at 840 and 764 cm^{-1} , demarcated by the dotted lines, which refer to the P and a phases respectively, in agreement with the literature. It was therefore possible to calculate the percentage of the P phase in relation to the a phase contained in the PVDF matrix.

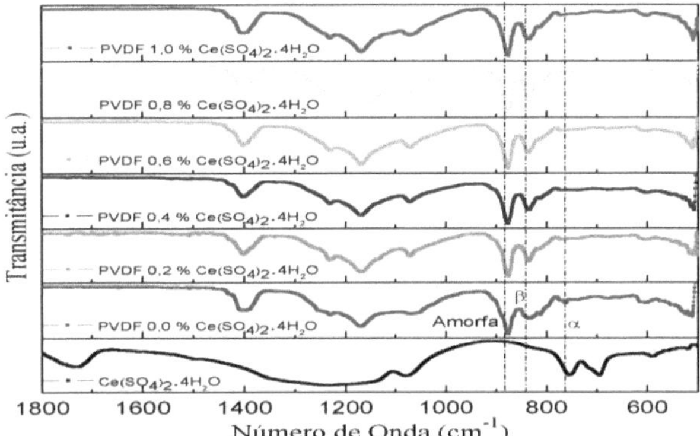

Figure 25 - FT-IR measurements of the PVDF/Ce(so4)2 .4H2 O samples at concentrations of 0.0; 0.2; 0.4; 0.6; 0.8; and 1.0 % cerium sulphate by mass, respectively.

The model proposed by Salimin and Yousef [68,69] was used to calculate the percentages of the P phase relative to the a phase, the results of which are shown in Figure 26. According to the results presented, it can be seen that the addition of cerium sulphate to the polymer matrix produces a significant increase in the relative percentage of P phase, with a variation of approximately 25% from the sample without cerium sulphate (pure PVDF) to the sample with 1%, i.e. the addition of cerium sulphate induces the formation of a greater amount of polar phase in the material.

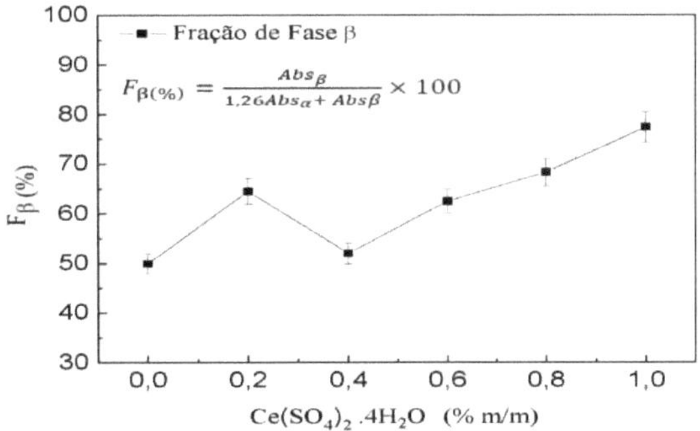

Figure 26 - Percentage of beta phase formation in relation to alpha, in samples of PVDF/Ce(so4)2.4H2O.

Dielectric constant measurements - PVDF/Ce(so4)2.4H2O

Considering that the addition of cerium sulphate induces an increase in the percentage of the P phase of the PVDF and that this is the ferroelectric phase, dielectric constant (s') measurements were carried out to check the behavior of this parameter as a function of the concentration of cerium sulphate and the frequency up to 18 KHz, these results are shown in figure 27.

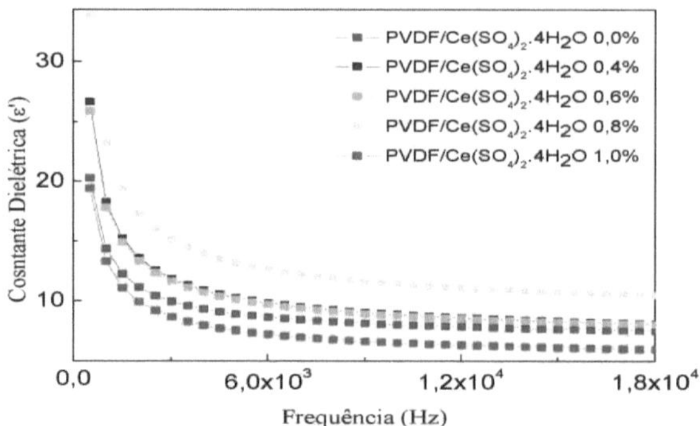

Figure 27 - Dielectric constant measurements for PVDF samples as a function of concentration and frequency up to 18 KHz.

For a better understanding of the results obtained for the s' measurements, a frequency of 1 KHz was chosen and the results are shown in Figure 28. As can be seen, for this frequency the value obtained for the pure PVDF sample is 14.3. Campos et al. reported that the average value they found was 13 [78], while Rozana et al. reported a value of 10 for the same frequency of 1 KHz [79]. As can be seen (Figure 28 (a)), the addition of cerium sulphate produces a significant increase of approximately 62% in the value of the dielectric constant at 1 KHz up to a concentration of 0.8% in the polymer matrix, but for a concentration of 1% the value of s' drops to approximately 13. This behaviour may be indicating a saturation condition for the 1% sample and may perhaps be inducing interfacial polarization (or the Maxwell-Wagner effect), which occurs at the interface of different materials, due to the different conductivities and permittivities of these regions, with the formation of space charges at the interfaces of these regions [80]. For a better understanding of the factor responsible for the effect, a graph was made of the dielectric constant values as a function of the relative percentage of P phase (Figure 28 (b)). The results obtained show the same behavior as the measurements as a function of concentration, i.e. the increase observed in the dielectric constant values is an effect caused by the addition of the dopant, since it causes an increase in the percentage of P phase.

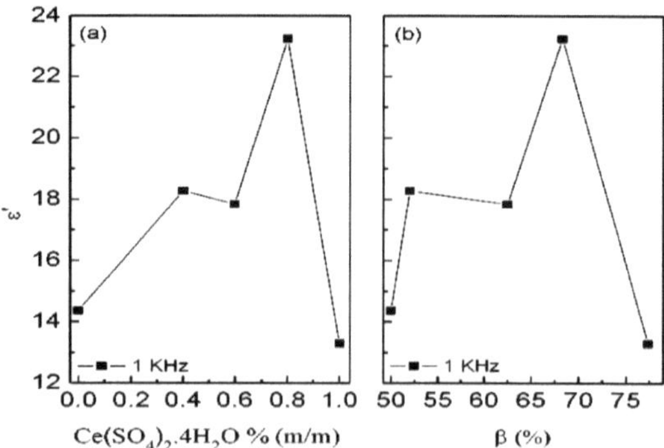

Figure 28- Dielectric constant measurements as a function of the concentration of Ce(so4)$_2$ .4H O$_2$ (a), and as a function of the relative percentage of P phase (b).

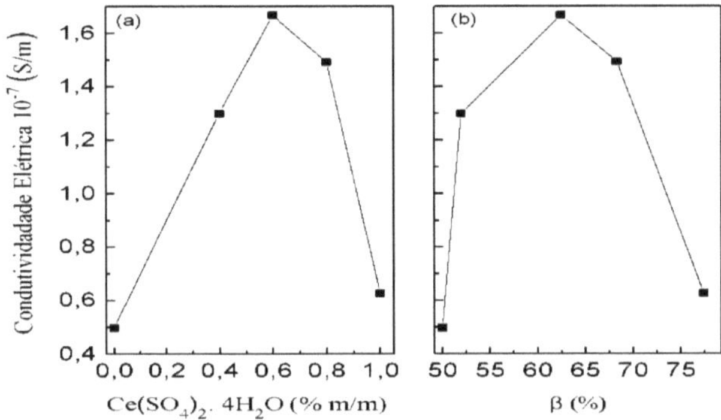

Figure 29- Electrical conductivity measurements as
a function of the Ce(so4)₂ .4H₂ O concentrate
(a) and as a function of the relative percentage of the P phase (b).

For the electrical conductivity measurements of PVDF, a value of 0.5 x10^{-7} S/m was obtained. In the literature, the reported value is 10^{-15} S/m, but depending on the methods used, the electrical conductivity values for PVDF can vary greatly from those reported in the literature [81]. As can be seen in Figure 29, the behavior is the same as the dielectric constant measurements. As the concentration of dopant in the polymer matrix increases, the value for conductivity increases to approximately 1.7 x10^{-7} S/m for the sample with 0.6 % Ce(so4)2.4H2O and from this value it decreases to 0.6x10^{7} (S/m) for the sample with 1 % dopant. This behavior suggests a relationship with the increase in dopant concentration in the polymer matrix.

5.1.2 UV-Vis and Fluorescence measurements - PVDF/Ce(so4)2.4HiO

UV-Vis absorption measurements are a tool used to study the band structure and electronic properties of doped and undoped polymers. It provides valuable information on electronic transitions from the valence band to the conduction band. Transition is direct when the wave vector for the electrons remains unchanged, but in the case of indirect transition, there is interaction with a vibration in the structure (phonons). In the present work, UV-Vis absorption measurements were carried out to verify the presence and effects caused by the addition of Ce(so4)2.4H2O to the polymer matrix. As can be seen (Figure 30), the absorbance spectrum of PVDF does not show a well-defined absorption band above 200 nm. The Ce(so4)2 sample shows absorption peaks at 225 nm, 240 nm and 320 nm, values which are close to those reported in the literature. Basha et al. identified a peak at 290 nm to which they attributed the 5d electronic transition$\rightarrow$ 4f [82]. The addition of cerium sulphate causes an increase in the peak

normalized absorption of the material, in the region between 210 and 400 nm up to a concentration of 0.6% Ce(so4)2 and a decrease in the absorption spectrum for concentrations of 0.8% and 1.0% for the same region. Figure 30 shows two absorption peaks, one at 250 and the other at 300 nm, due to the addition of the dopant. The behavior shown by the normalized absorption is the inverse of that observed in the electrical conductivity measurements, in which this parameter increases up to a concentration of 0.6 % of added dopant and then decreases for higher concentrations. From a practical point of view, the results show a significant change in the absorption spectrum of PVDF caused by the addition of the dopant.

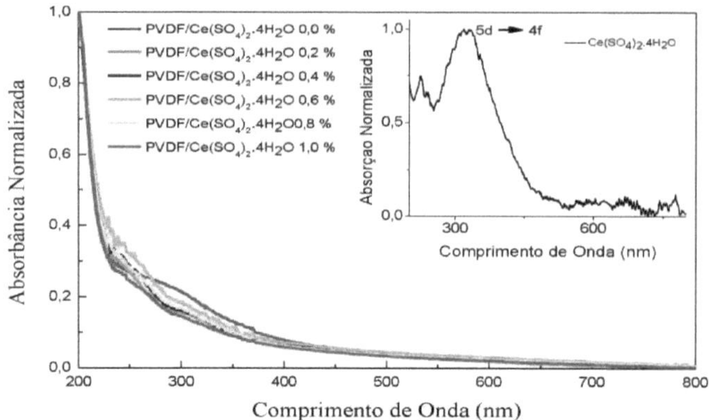

Figure 30- Absorbance measurements for the PVDF samples as a function of the concentration of Ce(SO4)2.4H2O.

Fluorescence measurements were carried out to check the influence of the addition of cerium sulphate on the natural emission of PVDF. In order to better understand the effects caused by the addition of cerium sulphate to the polymer matrix, a graph was made showing only the intensities as a function of the dopant concentration. However, it is known that interpreting fluorescence intensity measurements is difficult to correlate, as the intensity depends greatly on the amount of fluorophores. Although the samples under study have approximately the same thickness (60 p.m), they only differ by a few tenths of a micrometer. As can be seen in Figure 31, the PVDF without dopant shows a value of 500 for the maximum fluorescence peak, when the 0.2% dopant is added the intensity drops to approximately 345 and as the dopant concentration increases the intensity of the peak increases to 1154 for the 0.8% sample and then drops to a value of 670 for the sample with 1.0% dopant. This behavior shown in the fluorescence measurements is the same as that observed in the dielectric constant measurements, suggesting a saturation condition, where the sample begins to lose energy through non-radiative processes, causing the fluorescence to decrease. This result is interesting from the point of view of optical applications, as it provides information on the maximum amount of Ce(SO4)2.4H2O that can be added to the PVDF matrix without affecting its optical and electrical properties.

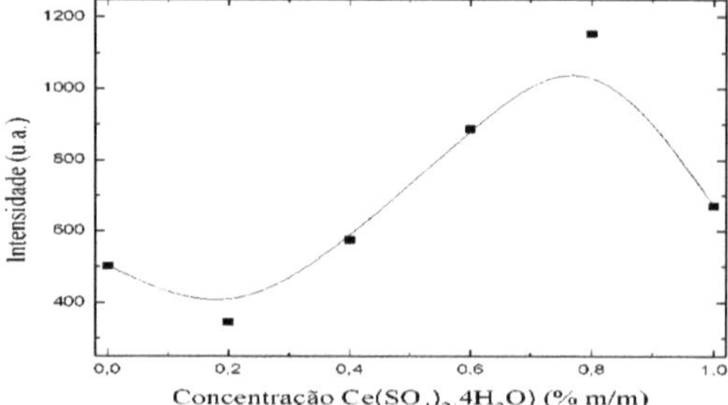

Figure 31- Maximum fluorescence of PVDF as a function of the concentration of Ce(SO4)2.4H2O.

The data obtained in Figure 32 shows that the addition of cerium sulphate not only causes a maximum increase in fluorescence intensity, but also shifts the peak of pure PVDF from 531 nm to approximately 503 nm for 0.8% PVDF/Ce(SO4)2.4H2O.

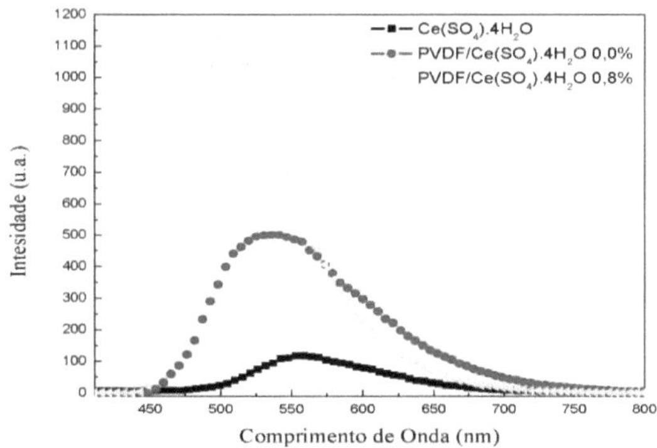

Figure 32- Fluorescence measurements for Ce(so4)2 .4H2 O, for the PVDF and PVDF/ Ce(so4)2.4H2O 0.8% samples.

    To better correlate the displacement of the peak of maximum fluorescence intensity for the PVDF samples, the fluorescence spectra were normalized and shown in figure 33. As can be seen, the cerium sulphate sample shows a half-height bandwidth of around 95 nm with a maximum peak at 560 nm, an effect that may be associated with the sulphate, since Ce^{4+} (the oxidation state of cerium in the compound) does not show fluorescence. The PVDF sample, meanwhile, has a half-height bandwidth of around 122 nm with a maximum intensity peak at 530 nm. Abdelaziz reports in the literature that the fluorescence band observed at 550 nm for the PVDF/PMMA (polymethylmethacrylate) sample must be associated with the physical rotation of the polar group (CH2-CF2) of the PVDF [83]. Figure 32 shows that as the concentration of the dopant in the polymer matrix increases, there is a shift to the left (region close to the cyan) of the peak of maximum intensity, whose peak detachment value for undoped PVDF in relation to PVDF/Ce(so4)2 1.0 % was 28 nm. For the 1.0% sample, a half-height bandwidth of approximately 82 nm was observed, with a maximum intensity at 502 nm. From the fluorescence measurements it can be seen that Ce^{4+} , even though it does not fluoresce as a sulphate, produces significant changes in the fluorescence spectrum of PVDF. The experimental results also indicate that further studies need to be carried out on PVDF/Ce(so4)2.4H2O samples in order to better understand the intrinsic effects of the matrix and dopant on the shift in the fluorescence peaks.

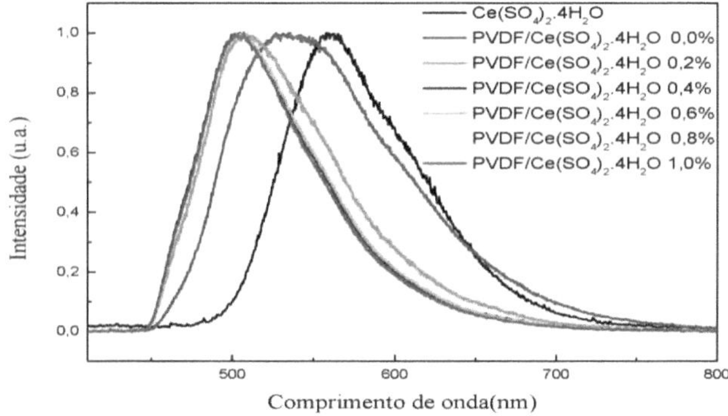

Figure 33- Normalized fluorescence spectra for the Ce(so4)2.4H2O, PVDF and PVDF/ Ce(so4)2.4H2O samples at different concentrations.

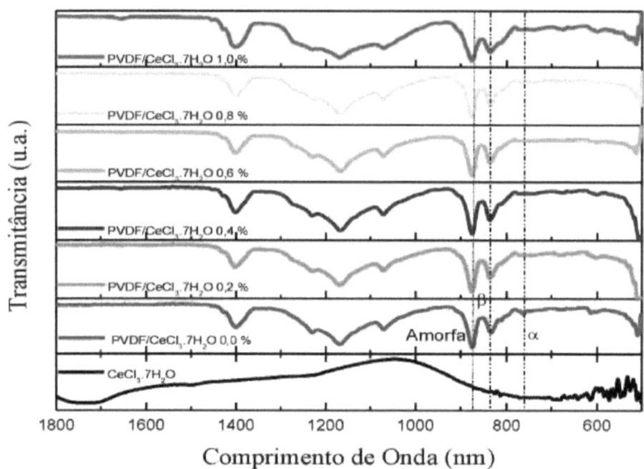

Figure 34- FT-IR measurements of the PVDF/CeCl3 .7H2O samples at concentrations of 0.0; 0.2; 0.4; 0.6; 0.8; and 1.0 % cerium chloride by mass respectively.

As previously discussed, the addition of dopants to the polymer matrix can produce changes in the material's structural properties. To verify these changes, FT-IR measurements were carried out on PVDF samples doped with CeCl3 .7H2O at concentrations of 0.0; 0.2; 0.4; 0.6; 0.8; and 1.0 % by mass. Abbreviated to CeCl3 to refer to CeCl3.7H2O. To check the influence of the dopant on the PVDF phases, the same methodology was used as for the PVDF/Ce(so4)2 samples [64,65]. The results of the relative percentages of P phase with the addition of CeCl3 induces an increase in the relative P phase from 68% to 82% up to the concentration of 0.4% CeCl3 added and decreases to a value of 54% for the 1% sample. Based on the FT-IR results, it is possible to suggest that increasing the concentration of CeCl3 induces a decrease in the relative percentage of P phase due to the interaction of the dopant with the polymer matrix, as shown in figure 35.

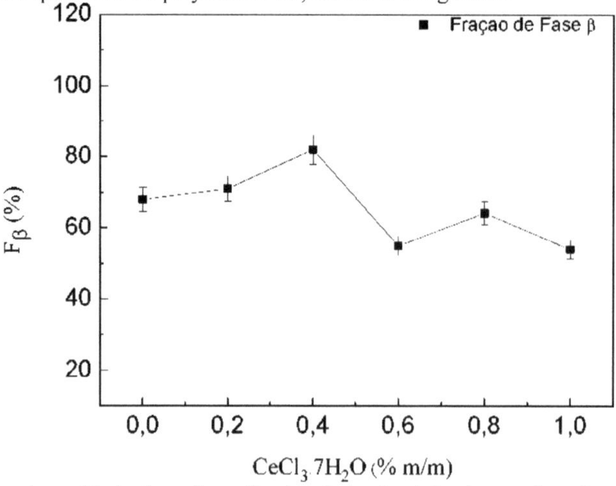

Figure 35 - Percentage of beta phase formation in relation to alpha, in samples of PVDF/CeCl37H2O.

Dielectric constant measurements - PVDF/CeCl3.7HiO

S' measurements were also carried out to check the influence of the dopant on the electrical properties of PVDF. Figure 36 shows the results obtained for the dielectric constant of the PVDF/CeCl samples₃ at all concentrations at a frequency of 18 KHz. And complementary measurements carried out at 1 KHz, the results of which are shown in Figure 37. As can be seen in Figure 37 (a), with the addition of cerium chloride the value of the dielectric constant varies from 13.9 to 9.8 up to a concentration of 0.4% CeCl3 and after this concentration it only increases to a limit value of 13.7 for the sample with 1% cecl3. However, considering an error of approximately 5 % (as shown in the error bar) it can be said with this result that there was no variation in the measured parameter disregarding the 0.4 % PVDF/CeCl3 sample, especially compared to other frequencies (Figure 36). For the dielectric constant measurements as a function of the crystalline phases (Figure 37 (b)) and considering the previous error, the values obtained indicate that it only decreases as a function of the relative P phase.

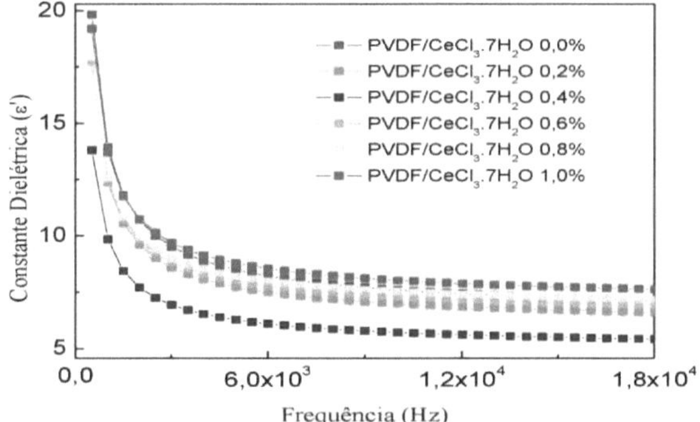

Figure 36- Dielectric constant measurements for PVDF samples as a function of concentration and frequency up to 18 KHz.

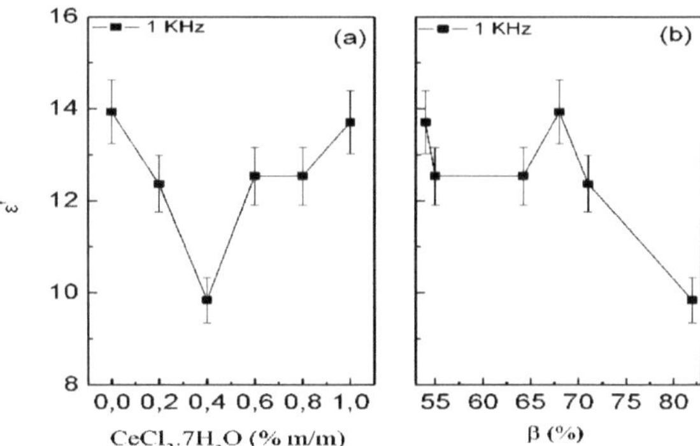

Figure 37 - Dielectric constant measurements as a function of the concentration of CeCl3 .7H2O (a) and as a function of the relative percentage of P phase (b).

The electrical conductivity of the PVDF samples doped with CeCl3 is shown in figure 38 (a). It can be seen that as the dopant concentration increases, the electrical conductivity value decreases from 0.0 - 0.4% as a function of the CeCl3 added to the PVDF matrix, and that from 0.4% onwards, the behavior of the samples suggests a steady state. However, according to the results shown in figure 38 (b), the electrical conductivity increases with the increase in the percentage of beta phase up to a limit of 70% and for higher percentages of beta phase the electrical conductivity value decreases. This concludes that the effects observed in the electrical parameters are caused by the addition of the dopant.

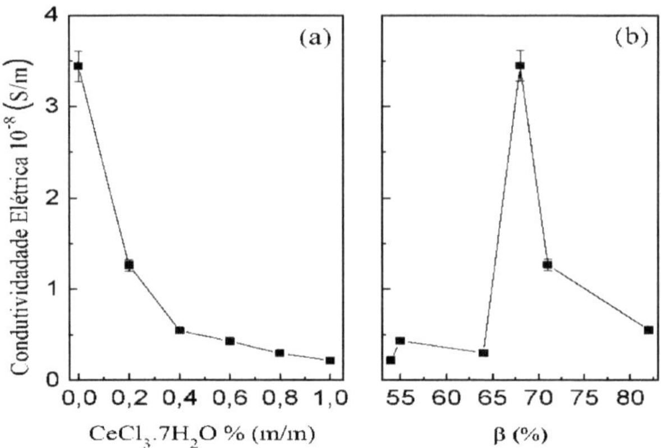

Figure 38 - Electrical conductivity measurements as a function of the concentration of CeCl3 .7H2O (a) and as a function of the relative percentage of P phase (b).

5.2.2 UV-Vis and Fluorescence measurements - PVDF/CeCkJHiO

These results are shown in figure 39 as a normalized spectrum for PVDF samples doped with CeCl3. The spectrum shows two absorption peaks at 253 and 303 nm which can be attributed to Ce^{+3} and a third peak attributed to Ce^{+4} [84,85]. In relation to these peaks, there is a decrease in absorption in the region between 200 and 600 nm. Inherent in the results presented, it can be seen from this spectrum that the dopant in the polymer matrix has a decrease in characteristic absorption, similar to the results presented for electrical conductivity.

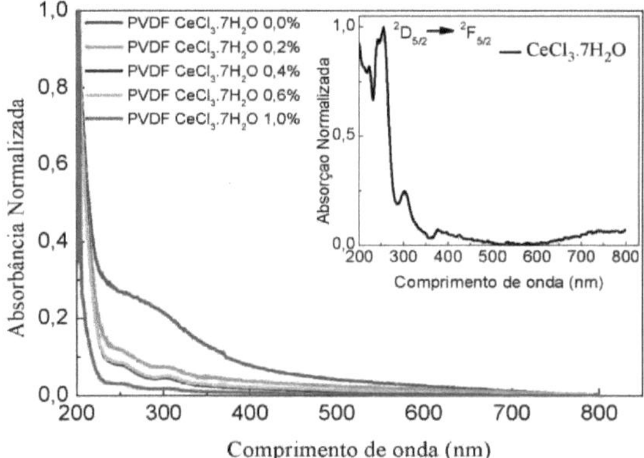

Figure 39 - Absorbance measurements for the PVDF samples as a function of the concentration of CeCl3.7H2O.

Figure 40 shows normalized fluorescence spectra for CeCl3, in which three characteristic fluorescence bands can be seen at 469, 557 and 596 nm. The addition of CeCl3 to the polymer matrix increases the bandwidth at half-height, from approximately 98 to 121 nm. In addition to the broadening of the fluorescence band, the increase in the concentration of the dopant causes a shift to the right of the peak of maximum intensity of pure PVDF by approximately 30 nm, up to a concentration of 0.8 % of added dopant, after which the observed behavior is the same as that of the sample without dopant. This behavior may be associated with a condition of material saturation, where the sample begins to lose energy through non-radiative processes. From the point of view of technological application, the addition of CeCl3 increases the natural fluorescence of PVDF, which could be used in devices for producing white light, due to the wide fluorescence range in the visible region of the electromagnetic spectrum.

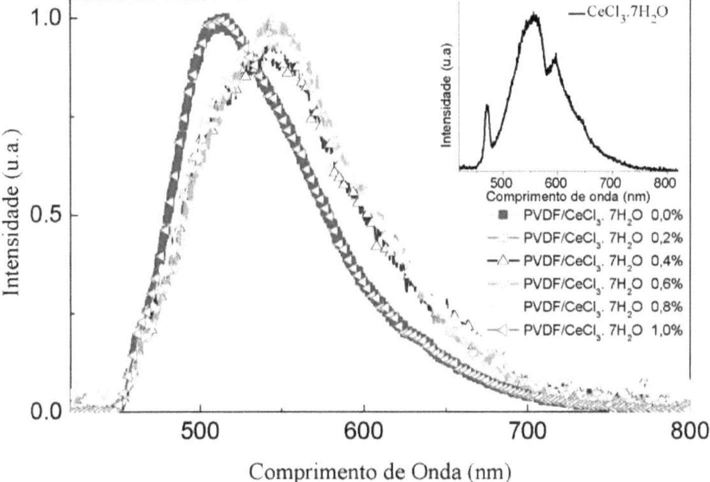

Figure 40- Normalized fluorescence spectra for the CeCl₃ .7H2O, PVDF and PVDF/CeCl₃ .7H2O samples at different concentrations.

Conclusions

In the present work, PVDF samples doped with cerium sulphate and cerium chloride were obtained and characterized for use in optical devices. The experimental results showed that the addition of cerium sulphate and cerium chloride to the polymer matrix caused significant changes in the relative percentage of the P phase in the PVDF samples. Given that the dielectric constant measurements showed different behaviors for the two dopants studied, increasing as a function of the cerium sulfate concentration up to the limit of 0.8% and decreasing in relation to the cerium chloride concentration up to the limit of 0.4%, indicating a possible condition of saturation of the polymer matrix. At the same time, the electrical conductivity and dielectric constant measurements showed the same behavior for the PVDF samples doped with cerium sulfate and a decrease in the electrical conductivity value for the PVDF samples doped with cerium chloride. The absorption measurements showed significant changes in the region between 210 and 400 nm for both the cerium sulphate and cerium chloride samples. In relation to the fluorescence measurements, it can be seen that cerium sulphate shifts the fluorescence peak to shorter wavelengths, as well as decreasing the width of the half-height fluorescence band and changing the fluorescence intensity, a behavior that may be associated with sulphate, since Ce^{4+} does not show fluorescence. It was also observed that the addition of cerium chloride shifts the peak of maximum fluorescence to longer wavelengths, as well as increasing the bandwidth at half-height, showing a wide range of fluorescence in the visible region of the electromagnetic spectrum.

In conclusion, the experimental results showed that the PVDF samples presented satisfactory results to be used as hosts for rare earth elements with potential technological applications. However, future studies need to be carried out with $PVDF/Ce(so4)_2$ and $PVDF/CeCl3$ samples in order to better understand the intrinsic effects of the matrix and dopant through the shift in the fluorescence peaks that these samples present.

Suggestions for future work

As a suggestion for continuing this work, the following topics will be highlighted:

TGA and DSC measurements were carried out to study the degradation processes that may occur in composites with temperature.

SEM and X-ray measurements were carried out for a more detailed characterization of the structure and morphology of these composites.

These studies will help to verify the feasibility of producing self-supporting films that can have a higher degree of crystallinity in the polymer matrix.

Scientific Production

1 - "Optimized Conditions of Crystallization in Solution for Obtaining P Phase Films of Poly (Vinylidene Fluoride)" (2011). SALMAZZO, G. R.; AMORIM, M. J. R.; FALCÂO, E. A.; CAIRES, A. R. L.; BOTERO, E. R. Paper presented at the 10th SBPMaT.

2 - "Synthesis and Characterization of Polyvinylidene Fluoride (PVDF) Cerium Sulfate" (2011). AMORIM, M. J. R.; SALMAZZO, G. R.; FALCÂO, E. A.; CAIRES, A. R. L.; BOTERO, E. R. Paper presented at the Internation Network for Advanced Multifunctional Materials.

3 - "Improvement in the Fluorescence Spectrum of the PVDF by Cerium Addition" (2013). AGUIAR, L. W.; AMORIM, M. J. R.; CAIRES, A. R. L.; BOTERO, E. R.; DOMINGUES, N. L. C.; CARVALHO, C. T.; FALCÂO, E. A. Paper presented at the XXXVI Encontro Nacional de Física da Matéria Condensada.

4 - "Synthesis and Characterization of Polyvinylidene Fluoride (PVDF) Cerium Doped" (2012).
FALCÂO, E. A.; AMORIM, M. J. R.; BOTERO, E. R.; CAIRES, A. R. L.; DOMINGUES, N. L. C.; RINALDI, A. W. ; CARVALHO, C. T. Paper presented at MS&T2012 - The Materials Science and Technology Conference, Pittsburgh.

References

FALCÂO, E. A. **Determination of the Optical and Thermal Properties of Telluride Glass and PLZT Ferroelectric Ceramics as a Function of Temperature and External Electric Field**, Maringá, 2006. Thesis (Doctorate in Science) - Department of Physics, State University of Maringá.

CANEVAROLO, J. S. V. **Ciência dos Polímeros**: Um Texto Básico Para Tecnólogos e Engenheiros. Sao Paulo: Artliber, 2004.

KAWAI, H. The Piezoelectricity of Poly (vinylidene Fluoride) **Journal of Applied Physics**, vol. 8, p. 975, 1969.

GREGORIO, J. R.; BORGES, D. S. Effect of Crystallization Rate on the Formation of the Polymorphs of Solution Cast Poly(Vinylidene Fluoride). **Journal of Applied Polymer**, vol. 49, n. 18, p. 4009-4016, 2008.

GREGORIO, J. R. Influence of Crystallization Conditions on the Morphology of Polyvinylidene Fluoride (PVDF) Films. **Polymers: Science and Technology**, p. 20-27, 1993.

DAVIES, G. T.; BROADHURST, M. G.; MCKINNEY, J. E.; COLLINS, R. E., **Journal of Applied Physics**. vol. 49, n.10, p. 4992, 1978.

KOCHERVINSKII, V. V. Piezoelectricity in Crystallizing Ferroelectric Polymers: Poly(vinylidene fluoride) and Its Copolymers (A Review). **Crystallography Reports**, vol. 48, n. 4, p. 649-675, 2003.

FUKADA, E. History and Recent Progress in Piezoelectric Polymers. **IEEE Ultrasonics, Ferroelectrics, and Frequency Control Society**, vol. 47, n. 6, p. 1277-1290, 2000.

FUKADA, E.; FURUKAWA, T. Piezoelectricity and Ferroelectricity in Polyvinylidene Fluoride. **Ultrasonics**, vol. 19, n. 1, p. 31-39, 1981.

CALLISTER, J. W. D. **Materials Science and Engineering**: An Introduction. 7. ed. New York: John Wiley & Sons, Inc, 2007.

PADILHA, A. F. **Engineering Materials**: Microstructure and Properties. 1. ed. Sao Paulo: Hemus, 2007.

CHUNHUA, X.; RUNPING, J.; CHUNFA, O.; XIA, W.; GUOYING, Y. Preparation and Optical Properties of Poly(Vinylidene Difluoride)/(Y0.97Eu0.03)$_2$O$_3$ rare-earth nanocomposite. **Optics Letters**, vol. 6, p. 763-766, 2008.

JUSTEL T.; NIKOL H.; RONDA C. **New Developments in the Field of Luminescent Materials for Lighting and Displays**. Angew. Chemie Int. vol. 37, p. 3084, 1998.

STEVENS, M. P. **Polymer Chemistry: An introduction**. New York: Oxford University Press, 1999.

YOUNG, R. J.; LOVELL, P. A. **Introduction to Polymers**. 2. ed, London: Chapman & Hall, 1994.

MANO, E. B.; MENDES, L. C. **Introdujo a Polímeros**. 2. ed., Sao Paulo: Edgard Blücher Ltda, 1999.

NICHOLSON, J. W. **The Chemistry of Polymers**, Cambridge: The Royal Society of Chemistry, 1994.

HEEGER, J. A. **Journal of Applied Physics**, vol. 105, p. 8475-8491, 2001.

MUNK, P. **Introduction to Macromolecular Science**, New York: John Wiley & Sons, Inc., 1989.

ODIAN, G. **Principles of Polymerization**. 3. ed., New York: John Wiley & Sons, Inc., 1991.

HIPPEL, A. R. V. **Dielectrics Materials and Applications**, London: Chapman & Hall, 1955.

KITTEL, C. **Introduction to Solid State Physics**. 8th edition, Rio de Janeiro: LTC, 2006.

LEWIS T. J. **The Piezoelectric Effect**. In: Conference on Electrical Insulation and Dielectric Phenomena (CEIDP), p. 717-720, Nashville, 2005.

SALMAZZO, G. R. **Synthesis and Characterization of Ferroelectric Composites**

Based on Poly (Vinylidene Fluoride)/(Pb, La)(Zr, Ti)O3 and Poly (Vinylidene Fluoride)/BaTiO3. Dourados, 2013. Dissertation (Master's in Chemistry) - Faculty of Exact Sciences and Technology, Federal University of Grande Dourados.

WONG, C. K.; SHIN, F. G. Effect of Electrical Conductivity on Poling and the Dielectric, Pyroelectric and Piezoelectric Properties of Ferroelectric 0-3 Composites. **Journal of Materials Science**, vol. 41, n. 1, p. 229-249, 2006.

FURUKAWA, T. Piezoelectricity and Pyroelectricity in Polymers. **IEEE Transactions on Electrical Insolation**, Wako, vol. 24, n. 3, p. 375-394, 1989.

MANO, E. B. **Polímeros como Materiais de Engenharia**, São Paulo: Edgard Blücher Ltda, 1991.

BERGMAN, J. G.; MCFEE, J. K.; CRANE, G. R. **Journal of Applied Physics**, vol.18, p. 203, 1971.

NAKAMURA, K.; WADA, Y. **Journal of Applied Polymer,** Sci. A- 29, p. 161, 1971.

SESSLER, G. M.; WEST, J. E. Applications, In SESSLER, G. M.; BROADHURST, M. G. **Electrets,** 2. ed., Berlin: Springer, 1987.

NALWA, H. S. **Ferroelectric Polymers**: Chemistry, Physics, and Applications, New York: Marcel Decker, 1995.

ODIAN, G. **Principles of Polymerization**. 4. ed., New York: John Wiley &Sons, 2004.

DROBNY, J.G. **Technology of Fluoropolymers**. 2. ed., Boca Raton, 2009.

LOVINGER, A. J. **Poly(Vinylidene Fluoride)**, In BASSET, D. C. Developments in Crystalline Polymers-1, 1982.

BROADHURST, M. G.; DAVIS, G. T. **Piezo and Pyroelectric Properties,** In SESSLER, G. M.; BROADHURS, M. G. Berlin: Springer, 1987.

ELLIOTT, S. R. **Physics of Amorphous Materials**. 2.ed. New York: John Wiley & Sons; Longman Scientific & Technical, 1990.

ULMAN, A. **An Introduction to Ultrathin Organic Film**: From Langmuir-Blodgett to Self-Assenbly. Academic Press: San Diego, 1991.

MICHAEL, P. C. **Langmuir-Blodgett Films**: An introduction. Cambridge University Press: New York, 1996.

HASEGAWA, R.; TAKAHASHI, Y.; CHATANI, Y; TADOKORO, H. **Journal of Applied Polymer,** vol.3, p. 600, 1972.

HASEGAWA, R.; KOBAYASHI, M.; TADOKORO, H. **Journal of Applied Polymer**, vol. 3, p. 591, 1972.

LOVINGER, A. J. In: BASSET, D.C. **Developments in Crystalline polymers**, London: Applied Science, p. 196-273, 1982.

LI-YING, T.; XIA, H.; O-BIN; XIAO-ZHEN, T. Study on Morphology Behavior of PVDF-Based Electrolytes. **Journal of Applied Polymer,** vol. 92, p. 3839- 3842, 2004.

NAEGELE, D.; YOON, D.Y.; BROADHURST, M.G. Formation of a New Crystal Form of Poly (vinylidene fluoride) under Electric Field. **Macromolecules**. vol. 11, p. 1297-1298, 1978.

WELTER, C. **Relaxor Behavior of Ferroelectric Polymers Subjected to Electromagnetic Radiation**, Belo Horizonte, 2008. Thesis (PhD in Physics) - Institute of Exact Sciences, Federal University of Minas Gerais.

STECKL, A. J.; ZAVADA, J. M. **MRS Bulletin**, vol. 16, 1999.

MAESTRO, P. Materials "Today and Tomorrow", Rhône-Poulenc: Paris, 1991

Greenwood, N. M.; Earnshaw, A.; **Chemistry of the Elements**, Pergamon Press: Great Britain, 1984.

SZABADVARY, F. Im **Handbook on the Physics and Chemistry of Rare Earths**; GSCHNEIDNER, Jr. K. A.; EYRING, L. Elsevier: Amsterdam, 1988.

MOELLER, T.; **The Chemistry of the Lanthanides**, Pergamon Texts in Comprehensive Inorganic Chemistry; Pergamon Press: New York, 1975.

PUCHE, R. S.; CARO, P.; **Rare Earths - Summer Courses at El Escorial**, Editorial

Complutense: Madrid, 1998.

KILBOUR, B. T.; **A Lanthanide Lanthology,** Molycorp, Inc., White Plains: New York, 1993.

MARTINS, T. S.; ISOLANI, P. C.; **Química Nova,** vol. 28, p.11, 2005.

CARRIJO, R. M. C.; ROMERO, J. R.; **Química Nova,** vol. 23, p. 331, 2000.

GARNER, J. P.; HEPPELL, P. S. J.; **Burns,** vol. 31, p. 539, 2005.

VICENTINI, G.; ZINNER, L. B.; Carvalho, L. R. F.; **Brochure - Produjo e Aplicares das Terras Raras,** Sao Paulo, 1980.

ABRÁO, A.; **Chemistry and Technology of Rare Earths**, CETEM/CNPq: Rio de Janeiro, 1994.

WANG Y. H.; ZHANG, J. C.; **Fuel,** vol. 84, p. 1926, 2005.

TANG, L. C.; YANG, T. G.; **Journal Chemistry physics**, vol. 1, p.18. 1995.

CALLISTER, W. D. J. **Materials Science and Engineering - An Introduction**. LTC: Rio de Janeiro, 2002.

KOMAMENI, S. **Nanocomposites. Journal of Materials Chemistry**, vol. 2, p. 1219, 1992.

GANG, X.; CHIEN, C.L. **Giant Magnetic Coercivity and Percolation Effects in Granular Fe-(SiO_2) Solids**. Applied Physics Letters vol. 51, p. 1280, 1987.

DAGANI, R. **Putting the" Nano" Into Composites**. Chemical & Engineering News vol. 7, p. 25, 1999.

SANCHEZ, C.; JULIAN, B.; BELLEVILLE, P.; POPALL, M. **Applications of Hybrid Organic-Inorganic Nanocomposites**. Journal of Materials Chemistry, vol. 15, p. 3559, 2005.

SANCHEZ, C.; ARRIBART, H.; GUILLE, M.M.V. **Biomimetism and Bioinspiration as Tools for the Design of Innovative Materials and Systems**. Nature Materials, vol. 4, p. 277, 2005.

GOMEZ-ROMERO, P. **Hybrid Organic-Inorganic Materials-in Search of Synergic Activity**. Advanced Materials, vol. 13, p. 163, 2001.

MAMMERI, F.; LE BOURHIS, E.; ROSES, L.; SANCHEZ, C. **Mechanical properties of hybrid organic-inorganic materials**. Journal of Materials Chemistry, vol. 15, p. 3787, 2005.

ALEXANDRE, M.; DUBOIS, P. **Polymer-layered Silicate Nanocomposites: Preparation, Properties and Uses of a New Class of Materials**. Materials Science & Engineering R-Reports, vol. 28, p.1, 2000.

SALIMI, A.; YOUSEFI, A. A. Analysis Method: FTIR Studies of P-phase Crystal Formation in Stretched PVDF Films. **Polymer Testing**, vol. 22, p. 699-704, 2003.

SALIMI, A.; YOUSEFI, A. A. Conformational Changes and Phase Transformation Mechanisms in PVDF Solution-Cast Films. **Journal of Polymer Science: Part B: Polymer Physics**, vol. 42, p. 3487-3495, 2004.

LAKOWICZ, J. R.; GRYCZYNSKI, I. In **Topics in Fluorescence Spectroscopy**, vol. 1, LAKOWICZ, J. R. Plenum Press: New York, 1991.

DONALD, J. R. **"Impedance Spectroscopy - Enphasizing solid materials and system"** John Wiley & Sons, 1987.

BACHMANN, M. A.; LANDO, J. B. A Reexamination of the Crystal Structure of Phase II of Poly (vinylidene fluoride). **Macromolecules**, vol. 14, p. 40-46, 1981.

COSTA, L. M. M.; BRETAS, R. E. S.; GREGORIO, JR., R. Characterization of PVDF-P Films Obtained by Different Techniques. **Polímeros: Ciencia e Tecnologia**, vol. 19, n. 3, p. 183-189, 2009.

TASHIRO, K.; KOBAYASHI, M.; TADOKORO, H. Vibrational Spectra and Disorder Transition of Poly (vinylidene fluoride) Form III. **Macromolecules**, vol. 14, n. 6, p. 17571764, 1981.

BORMASHENKI, Y. Vibrational spectrum of PVDF and its interpretation. **Polymer**

Testing, vol. 23, p. 791-796, 2004.

GREGORIO, J.; CESTARI, M. Effect of crystallization on the crystalline phase content and morphology of poly(vinylidene fluoride). **Journal of Applied Polymer**, vol. 32, p. 859, 1994.

SHIGEYOSHI, O.; TADADAO, K. Electrical properties of form III poly(vinylidene fluoride). **Ferroelectrics**, vol. 32, p. 1-11, 1981.

CAMPOS, J. S. C.; SANTARINE, G. **Eclética. Química**, Vol. 22, São Paulo, 1997.

ROZANA, M. D.; ARSHADL, A. N.; WAHIDL, M. H.; HABIBAH, Z.; ISMAIL, L. N.; SARIP, M. N.; RUSOP, M. Dielectric constant of PVDF/MgO nanocomposites thin films; **IEEE Symposium on Business, Engineering and Industrial Applications**, p. 18-22, 2012.

ROBERT, R.; KOWALSKI, E. L.; GOMES, D. M. G. Absorption and reabsorption current in dielectrics, **Revista Brasileira de Ensino de Física,** Vol. 30, p. 3307, 2008.

MATTOSO, L. H. C. **Synthesis, Characterization and Processing of Polyaniline and its Derivatives. Sao Carlos,** 1993. Thesis (Doctorate in Materials Science)   - Department of Materials Science and Engineering, Federal University of São Carlos.

BASHA, M. A. F. Spectroscopic, Magnetic, and Optical Characterization of Nanocomposite Films of Polyvinylpyrrolidone Doped with Cerium Disulphate; **Journal of Applied Polymer,** Vol. 122, p. 2121-2129, 2011.

ABDELAZIZ, M. Investigations on optical and dielectric properties of PVDF/PMMA blend doped with mixed samarium and nickel chlorides; **Journal of Materials**, 2013.

JACKSON, S, Photoionization of Ce^{3+}. **Journal Chemistry Physics.** Vol. 35, p. 844, 1961.

VABSON, G. B. **Characterization of Vitreous Systems Based on B2O3 and P2O5 Formers Doped with** CeO_2**.** Ilha Solteira, 2006. Dissertation (Master's Degree in Condensed Matter Physics) - Postgraduate Program in Materials Science, Universidade Estadual Paulista Julho de Mesquita Filho.

I want morebooks!

Buy your books fast and straightforward online - at one of world's fastest growing online book stores! Environmentally sound due to Print-on-Demand technologies.

Buy your books online at
www.morebooks.shop

Kaufen Sie Ihre Bücher schnell und unkompliziert online – auf einer der am schnellsten wachsenden Buchhandelsplattformen weltweit! Dank Print-On-Demand umwelt- und ressourcenschonend produziert.

Bücher schneller online kaufen
www.morebooks.shop

Printed by Books on Demand GmbH, Norderstedt / Germany